ADDITIONAL PRAISE FOR
THE ROBOT IN THE NEXT CUBICLE

"Today, leading enterprises are deploying robotic and intelligent process automation in the front and back office, more simply described as software robots. These bots are capable of handling work tasks previously offshored or realigned to shared service centers. *The Robot in the Next Cubicle* is a resource to help readers become more comfortable with this emerging relationship between humans and intelligent machines, and the evolving opportunity for increased collaboration, leading to deeper symbiotic interconnections."

—Mike Quindazzi, managing director, PricewaterhouseCoopers

"An examination of the rise of artificial intelligence and the role of humans in the future workplace. While he clearly depicts the rise of automation, he also gives readers a refreshing guide on how to leverage their social skills to maintain relevancy and sanity."

—Dan Schawbel, author of *Back to Human*, *Promote Yourself*, and *Me 2.0*

"Building a brand, whether for yourself or a business, takes time. This book is a call to action to start building your brand so you can succeed rather than be disrupted in the coming technological revolution."

—Brian Smith, UGG founder, and author of *The Birth of a Brand: Launching Your Entrepreneurial Passion and Soul*

"Some of the greatest opportunities for success came during times of economic disruption. *The Robot in the Next Cubicle* sheds light on emerging technologies and shows us how people navigated past economic upheavals to achieve their success. No one said it would be easy, but knowing what's coming, what to look for, and what you can do will make a big difference."

—Dr. Greg S. Reid, author of the Think and Grow Rich series

THE
ROBOT IN THE
NEXT CUBICLE

THE
ROBOT IN THE
NEXT CUBICLE

What You Need to Know to **ADAPT** and **SUCCEED** in the **AUTOMATION AGE**

LARRY BOYER

Prometheus Books

59 John Glenn Drive
Amherst, New York 14228

Published 2018 by Prometheus Books

Cover design by Liz Mills
Cover image © Suwin/Shutterstock (robots and office)
Cover image © Mavo/Shutterstock (man)
Cover image © Maksvil/Shutterstock (woman)
Cover design © Prometheus Books

Inquiries should be addressed to
Prometheus Books
59 John Glenn Drive
Amherst, New York 14228
VOICE: 716–691–0133 • FAX: 716–691–0137
WWW.PROMETHEUSBOOKS.COM

22 21 20 19 18   5 4 3 2 1

Library of Congress Cataloging-in-Publication Data

Identifiers: ISBN 9781633884106 (ebook) | ISBN 9781633884090 (pbk.)

Printed in the United States of America

To Hua, Lawry, and Max for their unending support and inspiration.

To my mother and father, whose love and wisdom continues to surprise and influence me. May they rest in eternal peace.

To working people everywhere. You deserve to understand the changes, challenges, and opportunities ahead and what you can do today to be ready for change and disruption. You and your families deserve no less.

CONTENTS

ACKNOWLEDGMENTS

This book is the culmination of a lifetime of experiences, education, and learning from people around the world. I am grateful to everyone who contributed to my ongoing education.

I'd like to acknowledge my editor, Steven L. Mitchell, and the entire team at Prometheus Books. Your belief in the concept and support at each step of the way have been tremendous. Thank you for helping bring the information and messages here to a wide audience.

For help with the research, feedback, and editing of the text I am grateful for the efforts of Stephanie Clarke at Clarke International Writing Services™. Bonnie Karpay, Lawry Boyer, Susan Rooks of Grammar Goddess Communications™, and Neil C. Hughes at Tech Blog Writer LTD™ for their reviews and support. Raoul Davis and Kirill Storch at Ascendant Group® for their long-time support with this project and its supporting projects to help guide people through the coming changes. A special thank-you to Leticia Gomez, my book agent, for finding the right home at Prometheus for this book.

Thank you to Kunal Sood, Sony Mordechai, and Ryan Long for inviting me to participate in the Novus Summit™ at the United Nations in 2016. It was there that the ideas for this book crystallized in my mind, inspired by the incredible works of so many people making an exponential difference. I hope this work helps achieve the UN Sustainable Development Goals.

Greg S. Reid and the Secret Knock® community and Berny Dohrmann and the CEO Space International® community have been a tremendous influence for my personal development and mind-set, showing me what's possible.

I'm eternally grateful to Wayne Buckhout, my mentor, teacher, and coach. Your unrelenting support and belief in me in my teens built my confidence that you can achieve anything you put your mind to. Not only that, you led by example, showing how to take charge of your life when you left a "safe" teaching job to build and expand the Cat's Cradle Country Shoppe® with your wife, Sandi. I still enjoy visiting the two of you there.

ACKNOWLEDGMENTS

Dr. Dale Ballou took this budding physicist under his wing in the economics department at UMass and helped direct and shape the blending of science, economics, and business. Dr. Ira Gang, who was my advisor at Rutgers University, unwittingly sparked an interest in labor and development economics. Dr. Bayard Catron, my advisor at the George Washington University, inspired me further with issues of global policy, ethics, and decision-making.

To my friends, colleagues, and trainers at iPEC®, especially Bruce D. Schneider, Luke Iorio, Mark Schall, Alan Samuel Cohen, and Francine Kolacz Carter, who helped me raise my own self-awareness and learn how to support others through the coaching process. To William Arruda, the Reach Personal Branding® community, and Dan Schawbel for shining the light on the value of personal branding.

To my group of close friends and supporters I've met on social media platforms Ecademy®, LinkedIn®, and beBee®—Thomas and Penny Power, Juan Imaz, Javier Cámara Rica, John White, Sarah Elkins, Neil Hughes, Marietta Gentles Crawford, Brigette Hyacinth, Milos Djukic, David B. Grinberg, Qamar Ali Khan, Prakashan B. V., Donna-Luisa Eversley, Lisa Gallagher, Susan Rooks, Robyn D. Shulman, Paul Drury, Arnie McKinnis, Jacqui Genow, Andrew Brooks, Chris Spurvey, Lynda Spiegel, Karthik Rajan, Jason Versey, Jan Barbosa, Syed Rashid Jamal, Kanwal Masroor, and Joe Kwon—thank you so much.

Finally, a big thank-you and *I love you* to my wife, Hua Yin, whose support at home has both inspired me and made it possible to dedicate the time to making this book a reality.

INTRODUCTION

My father graduated from high school, joined the US Army, learned a trade, and saw the world. When he left the army, he was hired by AT&T®, where he worked until he retired thirty-five or so years later. While we commonly understand that career model no longer works today, what I find helpful now is how my father managed to survive the waves of disruption during what we now know as the third industrial revolution, which consisted of the wave of transformations that led to computerization, electronics, and telecommunications technology. I didn't see it growing up, and unless you've given thought to it, you may not have noticed it either if you were raised between the 1960s and 1990s.

My father was a lineman, laying and repairing the copper wires that connected homes and businesses so we could talk on the phone. Back then, it was simply *the phone*, no need to call it a *landline*, and AT&T was *the phone company*. For much of the 1960s and 1970s, it hovered around number ten on *Fortune's* list of largest American companies.[1] There is no doubt that AT&T was a huge driver of the third industrial revolution, with its vaunted Bell Labs® driving disruptive innovations in everything from microwave communications to video calling. In that respect, it was great to be part of such a company creating the future. However, it wasn't all rainbows and unicorns. The new technologies being developed changed the nature of the work during that time, resulting in thousands of employees losing their jobs to the tech.

It wasn't all about technological change either. Employees lost their jobs and faced pressures for other reasons too—economic cycles, competitive pressures, and the eventual regulatory breakup of AT&T. The economy, political and social disruption, and life events continued as well. This was the time of economically "impossible" stagflation—simultaneous high unemployment and high inflation. The oil embargos. Cold War and détente. Iran hostage crisis. Savings and loan collapse. Relatives getting sick and passing away. It's a stark reminder that while technology marches forward, creating new jobs and destroying old ones, at the same time jobs are also created and destroyed by economics, politics, and other business factors.

I remember my father bringing home a laptop. That was the 1980s, while the rest of us were getting our first taste of the Apple II® (that's before the Macintosh®). He would work feverishly every night, even after I had gone to bed, working hard to understand what this machine could do and how it would help him keep his job and take care of his family. As copper wires were replaced with fiber optics, he'd bring home manuals and samples on which to work and practice. Repairing copper wires and glass fiber, while sounding the same, involved different skills. I remember him telling my brother and me that even when you graduate and are done with school, your education can never stop. He wasn't someone who read *Think and Grow Rich* or listened to Earl Nightingale, either. This was life showing it to him. It was his devotion to his work and to his family.

These days, there is a lot of talk about artificial intelligence, robots, and other forms of automation, and how these will either bring great fortune or destroy life as we know it. Neither view has much practical value for anyone. Unless you are a head of state, such issues are largely distractions from what's most important for you—taking care of yourself and your family. It is important to know that there is a lot more change happening than just with AI and robots. We're in the early stages of a new industrial revolution that is likely going to be more disruptive than any of the changes we've seen since the first industrial revolution displaced craftspeople, artisans, and other skilled workers with machines that could be operated by unskilled workers at lower cost while offering quality products affordably to more consumers. These sweeping changes are occurring across industries as well as creating entirely new lines of business.

The boom in automation is rapidly spreading to other types of businesses. As technology advances, machines are becoming attractive alternatives to employees in a broadening selection of industries to perform such tasks as tracking inventory; taking fast-food orders; conducting museum tours; growing and harvesting produce; handling logistics; and finding, retrieving, and processing store orders. Automated systems are even in the process of effectively replacing lawyers, accountants, and medical personnel. The prospect of making an initial investment in automated machines and systems that require limited operational oversight and maintenance is exciting for those running companies and businesses. Costs of using automation are small compared to the continual process of advertising, hiring, training, and managing a host of employees, who tend to create a high turnover rate.

Will new technologies create or destroy more jobs? This is the topic of discussion among political leaders, corporate heads, tech enthusiasts, and economists around the world. While it's a good question, it is not the question that should concern you. The questions you need to answer are:

1. What will happen to my job and career as new technologies come online?
2. If I lose my job, will I be able to find another job or source of income?

It's often said that we made it through the previous industrial revolutions better off and stronger than before, and therefore we should be better off after this industrial revolution too. The problem with this thinking is that industrial revolutions typically last decades. That can well be the entire working life of a generation or two of workers struggling to hold jobs, compete with machines, and take care of their families while waiting for the benefits of the industrial revolution to appear. The question is not "Will society be better off a hundred years from now?" but "How will **you** get through it tomorrow and in the years to come?"

The Great Recession may be an indication of what to expect. The key characteristic of sweeping changes with an impact on entire industries is what we need to be watching. For example, the entire financial services sector was affected. Those who lost their jobs early may have been fortunate enough to find another job in the same or a related industry. However, as job losses quickly rose, there was no place to go if you lost your job. Long-term unemployment rose to levels unseen in the post–World War II era and still remains higher almost a decade later than at the peak of most recessions before it. As the economy recovered, there was little need for companies to rehire past employees because technology evolved to the point where fewer people could do the work.

Complicating the picture is that more than ever, we live in a connected global community that enables us to both enjoy goods and services from people around the world and experience new sources of competition for our jobs and business products. To add to the dilemma, the global population continues to grow. Although the rate of population growth has been slowing recently, there are still a large number of people being added to the planet each day, which equates to more competition for existing jobs as well as more possible customers. The global population has nearly doubled since the 1970s to the present level of 7.6 billion.[2] It is estimated that the world population will grow by 50 percent by the year 2100, reaching 11 billion people.[3] What will 3.5 billion more people do to earn a living? Youth unemployment and underemployment are already serious problems in many parts of the world.

This book explores the issues and seeks to answer the questions that actually matter to you. I also seek to provide you with the means to successfully navigate the changes and disruptions of the fourth industrial revolution so you benefit from the technology instead of being left behind. We look to the past to understand what happened and what lessons we may learn to help us today. We move on to look at the present and what the future is likely to bring, then close with a practical guide to follow.

Alexander Graham Bell once said, "Before anything else, preparation is the key to success."[4] If you have wondered about or are being threatened by the expansion of automation and how it will affect both your chosen industry and personal life, then *The Robot in the Next Cubicle: What You Need to Know to Adapt and Succeed in the Automation Age* will answer many of your questions and help you prepare for the coming transition into a world increasingly run by machines.

PART I

UNDERSTANDING THE TECHNOLOGICAL ADVANCEMENTS REDEFINING OUR WORLD

The technological innovations of the fourth industrial revolution are more than introducing new technologies; they are on the verge of fundamentally disrupting the way we live, work, save and invest, think, and socialize, and every other aspect of our lives. We must understand the scale of what's happening in order to understand our place in it.

CHAPTER 1

THE FOURTH INDUSTRIAL REVOLUTION: A PROGRESSION TOWARD PERFECTION

We are living in perhaps the most exciting times in all of human history. The technological advances we are witnessing today are giving birth to new industries that are producing devices, systems, and services that were once only reflected in the realm of science fiction and fantasy. Industries are being completely restructured to become better, faster, stronger, and safer. You no longer have to settle for something that is "close enough," because customization is reaching levels that provide you with exactly what you want or need. We are on the verge of releasing the potential of genetic enhancement, nanotechnology, and other technologies that will lead to curing many diseases and maybe even slowing the aging process itself. Such advances are due to discoveries in separate fields to produce these wonders. In the not so distant future, incredible visions of imagination such as robotic surgeons that keep us healthy, self-driving trucks that deliver our goods, and virtual worlds that entertain us after a long day will be commonplace. If ever there were a time that we were about to capture perfection, it is now—and the momentum is only increasing.

Since humans first learned to make fire with primitive tools, technology has frightened, confused, and excited people, bringing with it massive transformations of individual lives and society as a whole. The changes have been both positive and negative as each new wave of technological advancement crashed onto the shores of advancing generations eager to apply their creations. Some people have benefited greatly from mechanical inventions, while others have significantly lost, having their lives turned on end by the radical changes technology delivered. The bottom line is that while everyone has experienced, endured, and adapted to the shifts created by technology's power in an isolated and slowly evolving way, it has been nearly two hundred years since we have seen the scale of change and disruption we are about to witness as we charge headlong into the twenty-first century.

We are currently experiencing what has been dubbed the *fourth industrial revolution* by Klaus Schwab, the founder and CEO of World Economic Forum®. This fourth wave of technological advancement is riding the coattails of three previous technological surges. The reason these movements are considered "revolutions" is because each one is more than a simple evolution of a technology. Rather, they involve significant advancements in technologies across many disciplines that converge and have far-reaching implications that change the way businesses and economies operate as well as how people live and earn a living. These changes cause disruption to the status quo of society, often leading to political turmoil. Technological advancements that industries are scrambling to perfect and employ in our present time have been built on the foundations of the earlier industrial revolutions, converging previous inventions and ideas with new and modified discoveries. Before taking a look at our current technological revolution and how people need to adapt in order to flourish rather than get cast aside, it helps to understand the pillars on which it has been built. Typically, we look back at industrial revolutions and conclude that they were worth the effort because we are overall better off because of them. However, living through revolutionary beginnings is not easy. Even those people who benefit greatly from such times may still struggle, while others are devastated by those changes the rest of their lives.

THE FIRST INDUSTRIAL REVOLUTION

The first industrial revolution began in the late 1600s with the mass introduction of mechanical production equipment and accelerated as new forms of powering this equipment with steam were found in the late 1700s. This revolution and evolution continued at a breakneck pace until sometime between 1820 and 1840. It was the beginning of mechanical substitution for labor that had traditionally been performed by animals or humans. Many skills that society relied on to function were provided by artisans who spent their lives learning and perfecting their crafts and services, passing on their knowledge to apprentices. Suddenly, their jobs were replaced by machines that could produce more and faster for lower cost, and their lives were turned upside down. The Luddite Rebellion of 1811–1813 was a direct result of the clash between the value of people and the efficiency of machines. It was led by groups of skilled craftspeople who were rapidly losing their livelihoods to several forms of mechanical replacements. Around that time, professional wool finishers had nearly all been replaced by machines, directly threatening the livelihood and well-being of their families (the Luddite Rebellion is covered in more detail in chapter 7). It was both a battle of human versus machine and a struggle between laborers and the owners of capital.

Certain aspects of society were radically altered by such inventions. The textile industry, for example, benefited greatly from the implementation of power-driven cotton gins and weaving looms. The introduction of machines boosted the production and profitability of company owners while bankrupting entrepreneurs and small businesses. Those who worked in the factories faced long hours and sometimes dangerous conditions. The majority of society (around 80 percent) consisted of the working class—those who were employed to carry out required jobs to meet the needs of the whole. As machines began taking over traditional human roles, employees lost their bargaining power with company owners. As a result, the need for skilled craftspeople declined, and wealth was transferred from skilled labor to capital owners and, to a lesser extent, unskilled laborers, including children.

The problems arising from large-scale use of machines first showed up in Great Britain and in cities elsewhere in Europe where large, established populations had already formed. Mechanized factories and processes drew even more people from their countryside farms and villages to the cities in hopes of finding steady employment, modern luxuries, and better pay. Industrialization drew migrant laborers of nearly every background, expanding town populations and significantly shrinking the pool of skilled workers who operated and excelled at their trades. Certain skilled trades such as weaving, sewing, and threshing were almost completely replaced by machines. Londoner Henry Mayhew recorded his concern for the growing poor in a document entitled *London Labour and the London Poor*. He wrote, "There is barely sufficient work for the regular employment of half of our labourers, so that only 1,500,000 are fully and constantly employed, while 1,500,000 more are employed only half their time, and the remaining 1,500,000 remaining wholly unemployed, obtaining a day's work occasionally by the displacement of some of the others."[1] Mayhew acknowledged that part of the reason for such dismal employment results was "the displacement of human labour in some cases by machinery."[2]

Along with the new machines came a need for new sources and forms of materials, new methods of manufacturing, and new types of skills. The first industrial revolution produced entire cities that housed these activities and had to either be located near energy sources such as water bodies or coal mines, or have adequate transportation networks to supply them with distant resources. This led to a decline in public health conditions as workers packed into shanty-towns that lacked proper water delivery and waste removal. Many people died as a result, and many others starved as a direct or indirect effect of being displaced by machines. Revolutions also occurred as the masses rebelled against poor treatment and conditions.

The budding United States was almost entirely an economy of agricultural production until the machines began to arrive around the turn of the nineteenth century. However, the industrial revolution set its base in northern cities, leaving those states south of the Mason-Dixon Line caught behind the times as far as technology was concerned. A power struggle ensued between those who would control human slave labor and those who would control capital—the money to build machines and reap their rewards. This played a key role in the American Civil War, which began in 1861. The South, although an agricultural powerhouse supplying the world with such staples as cotton, lacked the North's more sophisticated means of harvesting, transporting, and manufacturing products. Southern plantations and farms were heavily dependent on slave and other forms of human labor, and Southern cities had access to a much smaller percentage of railroads and banking institutions compared to their Northern neighbors.

By the time the war broke out, Northern states were enjoying a boom of wealth, riches, and freedoms created by heavy industries that were driven by an increased use of machines, which allowed them to produce upward of 90 percent of the country's manufacturing output. Technology was being applied to various areas, which gave the North a significant advantage over the South. The Northern army was more favorably armed, producing 3,200 firearms for every hundred that the South produced. This was enabled by Northern innovations in machining and inventions. Northern factories also allowed greater production of textiles for blankets and uniforms, leather for saddles and holsters, and pig iron for heavy cannons and guns. The North was applying new machines in agricultural industry as well. Mechanized planters, threshers, and reapers allowed for faster planting and harvesting, which gave the North a production edge over the agriculturally dominant South. The advancing technology of the first industrial revolution also drew more immigrants to Northern cities because they preferred the advantages of modern machines over the outdated and sparse comforts of the vast farmland in the South.

Transportation was another key factor in the outcome of the war. A more widespread and sophisticated railroad system provided fast and efficient delivery of troops and supplies to the Northern war effort. Toward the end of the conflict, Gen. Ulysses S. Grant utilized the growing railway system as well as technological advancements in steamships to overpower General Lee's forces and systematically take control of a growing number of Southern cities. The combination of modernized factories, machines, and transportation systems, along with a greater influx of workers, gave the North a greater advantage over the South. This horrible war that pitted brother against brother had the positive aspect of vastly accelerating the modernization of the United States through technological advancements. By 1900, the fledgling country was one of the greatest industrial powers in the world.

Industrialization created a flood of new jobs and provided a great many opportunities and benefits to those employed in its wake. However, a large portion of the employees filling the budding mechanized factories and businesses endured harsh labor conditions, long working hours, and overbearing bosses, all foreign to workers' previously slow and flexible way of life on farms and in small villages, where they basically managed their own affairs. The benefits obtained by working with machines outweighed the poor, cramped living conditions and the lack of time and energy to enjoy them.

In order to deal with such large-scale issues that sprang up almost overnight, governments were restructured and strengthened, and laws were enacted that established trade unions and created environmental protection policies. Those societies that embraced and adjusted to this first industrial revolution experienced rapid development. New technologies quickly created a broad divide between industrial and nonindustrial economies. Industrial nations prospered as new, mechanically powered manufacturing and means of transportation allowed goods to be processed and shipped faster, in more abundance and at lesser expense than traditional methods, which lowered costs for many goods. Notably, the price of food dropped overall. which helped reduce malnutrition and disease, giving industrialized nations a stronger, healthier citizen base. Traditional family gatherings and social festivals were replaced by long, hard hours of work on the factory floor, but on the bright side, music halls, theaters, and sports venues to enjoy such increasingly popular games as cricket, soccer, and rugby sprang up in cities, leading to the basic structure of the metropolises we see today.

THE SECOND INDUSTRIAL REVOLUTION

The second industrial revolution, sometimes referred to as the Technology Revolution, occurred around 1870 and continued into the middle part of the twentieth century. Steel replaced iron, which allowed the production of larger, stronger structures, such as skyscrapers and bridges, as well as more efficient modes of transportation like locomotives and ships. The vulcanization of rubber and creation of the combustion engine led to vehicular transportation on a mass scale. Electricity harnessed through power stations and light bulbs increased production rates and allowed for longer hours of operation. The invention of the telegraph and telephone opened the door to globalization via rapid, real-time communication. Literacy also exploded as mechanized printing presses churned out material and increasing numbers of universities were founded. New business processes, like the assembly line, reshaped the way work was done, enabling workers to produce more, faster, and at lower costs than ever before.

This wave of technology had a great impact on the social, corporate, and political arenas, dramatically altering nearly every aspect of daily living. Mass immigration, which began during the first industrial revolution, continued as people moved from the country to take factory jobs in big cities. By the 1920s, most people in the United States were living in urban settings. Ironically, machines moved to rural areas, where they increasingly replaced human farmers in growing crops and raising livestock. Nearly 40 percent of the entire US labor force was employed on farms at the start of the twentieth century, but at its close, that number dropped below 3 percent.

Early on, workers continued to experience harsh working conditions, long hours, and low pay. Safety issues plagued growing factories as fires, accidents, and poor health took their toll. It was during this time that Upton Sinclair wrote his classic book *The Jungle*, about the harsh working conditions in the meatpacking plants of Chicago. Those who became ill or were fortunate enough to reach retirement age had to depend on savings, since such benefits as pensions, overtime pay, vacations, and health insurance were nonexistent. Wages were still quite low, so it normally took the monetary contribution of all family adults and sometimes the children to pay rent and buy food. Housing conditions also remained cramped, as large or multiple families took shelter together in small rooms and apartments. Although indoor plumbing began taking hold during this time, it was rare at first, which limited the water available to an overcrowded populace for such things as bathing and cleaning.

Like the first industrial revolution, the early days of the second industrial revolution saw a significant amount of strife. In the United States, the Pullman Strike shut down the majority of rail delivery and transportation in the spring of 1894, resulting in thirty workers being killed and causing damages totaling more than $80 million. The strike began when the Pullman Company®, a manufacturer of rail cars, ordered a reduction in wages as a result of declining business in the rail industry. Hardships were particularly exacerbated for those employees living in the company town on Chicago's South Side, where Pullman owned the houses and stores. While the company reduced the wages of its employees, it refused to lower rents and prices. Pullman also chose to pay shareholders dividends despite the downturn in business, which further angered workers.

The initial strike led to involvement by the American Railway Union in June of that year. Union members totaling 125,000 across twenty-seven states refused to work on or service trains that utilized Pullman cars, which significantly crippled rail transportation west of Detroit, Michigan. Finally, President Grover Cleveland intervened, which led to a court injunction ordering strikers to cease, giving as an excuse their interference with the delivery of mail via the US Postal Service®. The strikers refused to comply, and President Cleveland ordered the US Army to end the strike, resulting in violent clashes between rail workers and government forces prior to the strike ending.

Struggles between the workers and business owners, or capitalists, continued through this time, shaping society and politics. It was during the second industrial revolution that Marxism, socialism, and communism started gaining popularity, especially in areas where the distribution of wealth was particularly uneven. As the effect of the second industrial revolution continued to redistribute income and wealth at all levels of society, it increasingly became a hot-button issue used during the Russian Revolution and even toward the rise of Adolf Hitler. At this time in the United States, so-called robber barons rose to become captains of industry who were responsible for shaping virtually every aspect of the American economic and political scenes. These two philosophically different worldviews between communists and capitalists would go on to shape global politics and economics for decades to come.

Like the first industrial revolution, the second wave of technology led to improvements on a broad scale, only further reaching. The discovery and introduction of electricity, mass production, and division of labor all began to work to the advantage of the average citizen. The process of manufacturing items was fine-tuned, and output was increased. Specialized regions were established where companies utilizing these new technologies could operate as an interdependent community. Wages began to increase, allowing more people to afford the products being mass-produced. The trend took off in earnest in 1914 when the headlines applauded Henry Ford for increasing daily wages to $5, nearly doubling those of other companies. Doing so cut down on costly employee turnover and the need to train new employees and ended factory shutdowns resulting from a lack of qualified workers.

Other strides also became evident as the twentieth century marched forward. The number of women in the workforce increased to around 60 percent at the end of the twentieth century from a mere 19 percent at its beginning. Child labor laws were implemented, eliminating the 1.75 million children laborers under the age of fifteen recorded in the 1900 US Census. Other laws were also enacted that provided a minimum wage, healthcare, pensions, vacations, reduced work hours, mandatory overtime pay, and more. Governments were also transformed from those that focused on the nation-state to those that focused on citizens, who were now heavily involved in driving the engines of production, transportation, and trade.

Of course, the pursuit of higher technology wasn't all beneficial, especially when it came to conflicts between humans. By the time the second industrial revolution began with its new surge of innovations and inventions, Europe already had such a tremendous stockpile of weapons that European leaders were at a loss for what to do with them. Various wars along the way such as the Crimean and Russo-Turkish Wars allowed for this surplus of weapons to be used, but they also led to advances in warfare technology such as greater mobility, longer reach and accuracy, and more intense destructive power.

By the time World War I broke out in 1914, new military machines were being pumped out at a faster rate than at any time previously. Existing weapons such as pistols, rifles, and grenades were improved, and the types of weapons were expanded to include such lethally efficient items as chemical weapons, torpedoes, and machine guns. Mechanized vehicles were also a major addition to WWI battlefields, as technology improved other modes of mobile war machines. Armored cars, tracked transport vehicles, zeppelins, airplanes, and tanks were introduced, although most remained fairly clumsy or confined to minor roles. For example, planes were mostly used as observation vehicles to scout enemy movements or strategic targets, and it wasn't until the end of the war that tanks were fitted with revolving turrets.

World War II began in 1939 and not only saw the introduction of new and improved weaponry but also inspired even greater technological development in a wide array of fields. Communication technology was a major factor, as radios allowed better and more accurate strikes by both aircraft and ground weaponry. Handheld and hand-operated weapons were improved to include semiautomatic and then full automatic machine guns, bazookas, and mortars. Advanced development of armor made vehicles more resistant to the ravages of war, and airplanes were improved beyond a mere observational role to allow better fighting capability, including with machine guns and bombs, as well as to add important dimensions such as air transport and paratrooper drops. Technological advancements contributed to large seafaring fleets consisting of aircraft carriers and battleships that not only had massive firepower and troop/supply delivery capabilities but were also equipped with sonar and radar. Amphibious modes of operation were also successfully engaged as landing craft were developed, making sea-to-land assaults possible and effective.

Despite the expansion of military machines that allowed humans to more effectively and efficiently destroy and kill each other, the advancement of technology provided some equally dramatic benefits for those citizens seeking to live in peace and pursue a better life. A long list of useful items and devices were spawned from the war and applied to civilian life and the business world. Computers were developed, with their main purpose being to decipher Enigma codes. Penicillin was developed from the need to treat serious infections on the battlefield and has saved countless lives since. Jet engines, pressurized cabins, rockets, and satellites were also direct war contributions and have led to developments in travel speed and comfort, peacetime space exploration, and digital signaling used for advances in communication and vehicle navigation. Nuclear power was also developed as a means to quickly stop what was seen as a very costly war, both in resources and in lives, and has become a source of inexpensive energy. Other useful inventions that originated from the war effort include radar, microwaves (which led to those handy-dandy ovens), superglue, synthetic rubber, ballpoint pens, and photocopying.

Further mechanization of factories led to more effective and efficient assembly lines and more jobs. The result of all of this new technology was a greater reach of trade and communication, a rise of individual rights, and the expansion of a better-educated population. A larger segment of society was able to obtain more goods, which led to the culture of materialism. Although more people could experience luxury, comfort, and an overall increase in their standard of living, the boom of unregulated manufacturing practices set the stage for future environmental and resource issues. The prerequisites to global warming, climate change, and resource depletion that we are experiencing on a grand scale today all began in earnest during the second industrial revolution.

THE THIRD INDUSTRIAL REVOLUTION

The later part of the 1960s ushered in the third industrial revolution, or Digital Revolution, which inundated society with electronics, automated production, and computerized data technology. Such new developments eventually spurred the growth of globalization, which enabled further reductions in the costs of parts, labor, technology, and finished goods. Global production networks grew out of the movement that developed new manufacturing techniques. Countries left behind in the previous two revolutions increased development and began adding to the overall production of cheaper, more abundant, and better goods and services. During this time, advances appeared in industrial robotics, miniaturization, personal computers, portable audio and video recording, the internet, and fast-food production. Rising technology played a role in the rise and fall of Communist regimes; the Cold War; the Deming approach to manufacturing; and interplanetary space travel, which led to the moon landing and to probes that flew by all of the planets and landed on Venus, Mars, and even a comet. It continued to march forward, reshaping the world economically, politically, and socially. A major change was the way in which wars were fought. Instead of being carried out on a physical battlefield with destructive weapons, war was waged primarily with ideas that utilized the airwaves, internet, and cross-cultural exchanges to gain an advantage.

All of these trends contributed to the decline and even demise of many manufacturing facilities as the technology that made models more efficient was adapted. The oil and gas industries, in particular, took a nosedive as such energies began being replaced by more advanced energy processes that were environmentally friendly and clean. To compound the problems of the energy industry, entire infrastructures designed to transfer and utilize such fuels were aging and falling into disrepair, and noticeable shifts in weather and climate began to expose the cost of fossil fuel gluttony.

Advances in robotics were used by the Japanese to overtake US car manufacturers in the 1970s and 1980s. Enhancements in robotic assembly-line manufacturing—combined with Japanese workers who were hardworking, loyal, and focused on quality—gave the nation's car manufacturers a significant edge over US companies. Japanese workers were also treated much better and made to feel like they were important assets, which magnified their already-instilled good work habits. Cars produced by companies like Honda® and Toyota® became common sights on US roads due to their higher standard of quality, beating out American-made cars. The near-bankruptcy and subsequent government bailout of Chrysler® was due to strong Japanese and German car sales that drove the American automaker to expand operations in an attempt to compete. However, recessions, energy crises, and government-imposed regulations concerning fuel efficiency and environmental standards were straws that contributed to breaking the metaphorical back of Chrysler.

How US workers were treated was another important factor in the decline of American-made automobile quality and sales because although US companies were deploying robotics just as Japanese ones were, US workers remained divided between blue- and white-collar positions. This led to frequent union strikes and battles as disgruntled blue-collar workers fought for better organization and higher wages. As unions organized, grew more powerful, and won large increases in benefits and wages, auto manufacturers began moving to other countries where labor was both extensive and cheap. Tax breaks sweetened deals that enticed some automakers and their supporting businesses to foreign shores. Many companies were already in the process of adjusting business models and management structures to better compete in rapidly growing global markets that were feeling the effects of various pressures—including the expansion of a wide array of technologies.

The Digital Age opened up a large number of opportunities for global expansion, as business could now be conducted from practically any location on the planet that had access to increasingly large data-sharing systems. Major governmental shake-ups resulted as the Digital Revolution marched forward. The mass sharing of information led to the creation of broad and far-reaching free trade agreements as well as a mass exodus of jobs to offshore destinations. This led to job losses in the United States and certain European nations as jobs were transferred, but it also resulted in less expensive products and the expansion of consumerism, and it worked to strengthen developing nations that could house job transference by providing cheaper labor and operating costs. Once again, advanced technology directly resulted in revolutions that changed the face of the political map. Blossoming technology contributed to the fall of the Soviet

Union and Eastern Bloc, as well as to uprisings in China like the Tiananmen Square protests. Also, mobile devices and social media were contributing factors in the Arab Spring uprisings.

Undoubtedly, the arrival of computers, with the internet connecting them, drove massive changes as more industries incorporated computerized mechanisms. This new technology brought with it a plethora of both pros and cons, as did previous technological revolutions. On the positive side, factories and businesses became more efficient as computerized machines and software systems increasingly replaced a costly, faulty, and vulnerable human labor force. Machines took over some dangerous jobs and jobs that were repetitive and led to stress injuries. Fewer people were required to operate and maintain these systems that could produce more over longer periods of time at less expense, and new openings became increasingly available for those who acquired the specialized skills required to keep those systems operating and advancing unhindered. All these factors also led to lower prices, enabling more people to participate in the benefits these innovations had to offer.

It wasn't all good news, however. Many thousands of factory workers, telephone operators, secretaries, and others were displaced by digitization. This was also a factor in creating a widening gap between highly skilled, better-paid workers and low-skilled workers in low-paying positions. A widening gap between the wealthy owners of capital and the workers began to emerge.

Digitization has created vulnerabilities that didn't exist before. The Black Monday market crash of October 19, 1987, was driven by computerized trading. The Great Recession of 2007–2008, which saw the crash of the housing market, was also intensified by digitization, as bankers had access to further reaches of unregulated and unmonitored transactions. Hackers and terrorist groups have often taken advantage of weaknesses in computerized systems to wreak havoc on consumers, companies, and entire social centers. Digitalization has become a weapon that holds as much power as militarized machines, if not more.

THE FOURTH INDUSTRIAL REVOLUTION

The fourth industrial revolution is rising from a series of breakthroughs in data storage, connectivity, analytics, materials science, bioengineering, and more. In the words of Dr. Klaus Schwab, who has written a book and numerous articles on the subject, "We are at the beginning of a revolution that is fundamentally changing the way we live, work, and relate to one another. In its scale, scope and complexity, what I consider to be the fourth industrial revolution is unlike anything humankind has experienced before."[3]

Indeed, this revolution in technology seeks to solve various problems that have arisen from the previous three, thus improving human conditions. Advanced technology, particularly in the areas of mobile devices, social media sites, and other related sources, is being used to minimize conflict by better analyzing trends and providing real-time warnings and responses to emerging conflicts. This is due to more than half of the world's population either owning or having access to computers or mobile devices.[4] With such a large and growing number of people connected by technology, the potential for sharing information and finding peaceful means for solving problems is massive. The challenge is to remain focused on harnessing and utilizing rapid advancements in technology for the betterment of humanity. This age of high-tech development in such areas as biotechnology, nanotechnology, 3-D printing, the Internet of Things, quantum computing, big data analytics, energy production and storage, materials science, space industrialization, virtual reality, blockchain, drones, robotics, artificial intelligence, and more is so important that the January 2016 annual meeting of the World Economic Forum®, held in Davos-Klosters, Switzerland, had the theme "Mastering the Fourth Industrial Revolution." This topic took high precedence at the 2017 gathering of the WEF as well.

We are seeing evolved technology radically transform our lives as it is meshed with and becomes dominant in home appliances, computers, financial products, smart devices, data storage systems, transportation, communication, and more. Certain technologies are becoming so advanced that they are being merged with the human body in order to increase intelligence and strength, heal injuries, diagnose illnesses, enhance performance, instantly operate external machines, and perform other such tasks. Nanotechnology is allowing science to delve into the microcosm to accomplish such "miraculous" feats as curing disease, eliminating pollution, developing quantum computing, and creating miniscule power sources. Genomics is on target to revolutionize healthcare by identifying and then either treating or eliminating unwanted inherited traits ingrained in our genetic material. Genomic medicine is defined as "an emerging medical discipline that involves using genomic information about an individual as part of their clinical care (e.g., for diagnostic or therapeutic decision-making) and the health outcomes and policy implications of that clinical use."[5] This approach to medicine utilizes DNA sequencing and bioinformatics to analyze entire genome structures in order to identify their functions.

Facebook® has recently announced plans to develop an interface that allows brain-to-computer communication without the use of invasive implants. The goal is to provide a means by which thoughts can be scanned, detected, and then translated to text without the need for physical typing. The initial target is allow mental typing of one hundred words per minute via thought-only technology. The potential for such brain-computer interfacing may lead to the ability to control virtual reality and augmented reality devices without the use of external controllers and screens. Another project the Facebook research and development team is working on involves what is termed "skin-hearing," which is the ability for computer software to translate sound into frequencies that can be sent and received through the skin in much the same way as the ear. If successfully developed, such technology could allow people with hearing limitations or loss to effectively communicate.[6]

Billionaire technology developer Elon Musk is experimenting with a similar brain-computer interface known as Neuralink®. The idea is to create a device that can be implanted in the brain, which gives the receiver the ability to interface with computing devices in much the same way as artificial intelligence.[7] Musk told a crowd gathered at the World Government Summit® held in Dubai that "over time, I think we will probably see a closer merger of biological intelligence and digital intelligence." He believes this cyborg type of technology could be used as a means by which humans can effectively improve themselves.[8]

A restructuring of the world's financial system is yet another target of new technology as cryptocurrencies and blockchain technology expand in popularity and use. Blockchain is a distributed ledger, a recording of transactions that is stored in multiple databases, not just a single database. When new transactions are made, they are compared against the multiple records to ensure the records match and the transaction can be legitimately made. Cryptocurrency, such as Bitcoin, is one application of the blockchain. Bitcoin was the first digital cash phenomenon to be unveiled, in 2008; creator Satoshi Nakamoto described it as "a peer-to-peer electronic cash system."[9] Although Bitcoin got off to a slow start, the idea of using it and other virtual currencies has gained momentum. A September 2017 survey revealed that 78.5 percent of Americans had heard of Bitcoin, and 40 percent of those surveyed showed an openness to using it in the future. However, the currency's actual use is still lagging, as only 14 percent of Americans confessed to actually owning it.[10] Bitcoin continues to lead the pack of cryptocurrencies, but it is facing some serious competition from other virtual currencies that have arisen since its conception. Leading cryptocurrencies at the moment include Litecoin®, Ethereum®, Zcash®, Ripple®, Neo®, Cardano®, and Monero®, among others.

Blockchain is much more than cryptocurrencies. Applications are being developed to manage electrical grids, supply chains, and accounting systems as well as provide cybersecurity support and much more. Blockchain has the potential to revolutionize the way business and transactions are handled. Companies that develop and adopt this technology have greater opportunities to grow and displace or disrupt competing companies that do not. The combination of blockchain technologies and the quantum computing capabilities that are currently developing will likely have staggering impacts on a wide array of businesses and jobs. Businesses can utilize blockchain applications to organize healthcare data, create effective media platforms, track energy usage and transactions, and market real estate. The banking and finance industries will find blockchain useful for improving capital markets, offering peer-to-peer transactions, maintaining account encryptions that help deter and detect financial scams and money laundering, improving trade finance deals, and providing secure and transparent insurance policies and claims. Blockchain apps are also useful to governments for improving record management, tax gathering, regulatory oversight, voting platforms, and identity management. Blockchain technology also offers a variety of other benefits in such areas as business licensing; education; fraud prevention; securities movements; digital manufacturing; and tracing and tracking food, vehicles, cargos and various types of funds.

The lines between the real world and the virtual world of technology are becoming increasingly blurred as larger segments of society are inundated with the workings of machines. It is not individual technology that is creating this new industrial revolution, but the convergence of many technologies, new forms of energy generation and storage, and innovative business models, all of which together are being used to create something new and unpredictable. And because we are in the earliest stages of the fourth industrial revolution, we cannot yet see all of the new technologies that will be developed; know how they will be combined; or understand the impact they will have on society, jobs, the economy, and the way we live. We can, however, see that there are a great number of technologies being developed that are on the verge of being commercialized and becoming a real part of our lives. Deeper changes to the "norm" of business, the job market, government, and society as a whole are taking place at a lightning-fast pace. Companies, societies, and individuals who have already made major adjustments to compensate for the third industrial revolution are being forced to once again make cultural and organizational adjustments to meet newly surfacing requirements and challenges. The irony of the situation is that advances in high technology are being sought to solve the very problems being created by its ongoing development.

As with the three previous industrial revolutions, the present movement is causing societal as well as industrial disruption. Those who adapt are discovering the benefits of high technology, while those who do not are being left behind. What is being dubbed the Analytics Revolution is being hailed as an era of great promise. As technology advances, it will undoubtedly offer unprecedented opportunities if used responsibly. We have the opportunity to enjoy healthier, safer, and happier lives in a world with smart cities that are powered by natural sources of energy, offer products and services that are more efficiently produced and cost-effective, and are protected by cybersecurity systems.

However, the ongoing fusion between humans and robots will undoubtedly also produce massive and relatively unknown changes that will tempt governments, scientists, and corporate heads to cross lines of ethics, safety, and reason. This current industrial revolution has the potential to cause greater disruption than its predecessors because the displacement of human workers is projected to be higher than before. Study after study has revealed consistent data backing the loss of tens of millions of jobs.[11] This move toward massive displacement of both blue- and white-collar employees comes at a time when the world population is nearing eight billion. The mass move to high technology will ultimately place a much greater strain on company heads, governments, and other policy makers to find ways to deal with dramatic societal changes. We simply do not yet know the extent of those changes and how they will ultimately affect our individual and social lives. Nonetheless, it will be advantageous for people to prepare in order to ensure they come out better for the advances in technology, not left behind.

THE MOMENTUM OF CHANGE

There is a host of new technologies working to create dramatic changes in the present that will drastically affect economics in the not too distant future. This change is occurring in practically every segment of society and on a global scale. The explosion of shared information and digitally accessible data has made massive advancements in technology possible. Changes are happening so rapidly that individuals; social circles; and the programs, businesses, and economies they feed find it extremely taxing to keep up. Those who struggle and fall behind may very well fail in their efforts to survive.

Since the humble beginnings of humanity, technology has been considered a positive attribute. From more precise hunting weapons and better construction methods to enhanced forms of water delivery and agriculture, technological advancements simplified and improved the lives of ancient tribes, helping them grow into blooming, productive civilizations. The arrival of the Industrial Age saw an explosion in technology that has continued to expand. Over the span of not much more than a century, a mere tick on the universal clock of time, humans have created such "miracles" as cross-country and worldwide transportation, disease prevention and eradication, global real-time communication, and space exploration. As a result, we live longer, are healthier, have more money to buy what we need and want, and can take time to enjoy recreational activities or hobbies.

Although a great many positives rose out of it, the first industrial revolution wasn't an easy or celebrated walk in the park. It caused major disruptions in society that lasted for decades, longer than the working life of most people. Generations of workers and their families were affected by uncomfortable and even turbulent changes. People were suddenly dealt with differently, policies had to undergo major overhauls, and shifts in education and work schedules occurred, all of which disrupted governance, structures, guidelines, processes, and lives. Adaptation was vital in keeping up with the technological developments being implemented throughout society.

Subsequent industrial revolutions caused equally stressful disruptions. For example, the third industrial revolution brought an abrupt end to many thousands of telephone operator jobs. As with most positions radically transformed by new advancements in technology, those workers who were at the end of their career were able to retire without noticeable trauma. However, younger workers who lost their jobs had to seek new sources of expertise that fit the new paradigm, or they found themselves cast aside because they held knowledge and experience that was no longer relevant or needed. These changes occurred over shorter periods of time, which only magnified the stress placed on those affected.

There appears to be no slowdown to the advancement of technological applications. As a matter of fact, the speed at which ideas and inventions are manifesting is accelerating to a near-breakneck point. Robotics, nanotechnology, quantum computing, 3-D printing, artificial intelligence, and other rapidly advancing sciences are opening doors into unfamiliar areas. We are actually achieving the outcomes of such possibilities before we fully understand where they will take us. This practice of developing and initiating vastly unexplored technologies that have the potential to significantly alter our physical, biological, digital, and economic worlds is both extremely exhilarating and potentially dangerous. It is wise to step back and take a sober look at how these technological advancements will improve our lives, while at the same time taking into consideration the possible negative implications they will have on individuals, companies, social structures, and economies as a whole.

THE ARGUMENT: A BETTER LIFE

One strong argument for the development of new technologies is that they will continue the process of vastly improving our lives. For example, the field of robotics is revolutionizing the manufacturing industry, which is considered to be beneficial for the employer, employee, and customer. It is trumpeted by company heads that automated robotic machines not only save money, cut down on employee training and expense, and improve efficiency but also increase pay for specialized employee operation, decrease accidents and healthcare expenses, and reduce overall cost for customers. These claims are indeed true, and largely embraced by investors and others set to benefit from technological advancements.

However, there remains a major long-term problem that very few are talking about. Companies utilizing robotics reduce expense and boost profits due to one main outcome—fewer employees are needed to produce and deliver products. Of course, that is viewed as a superpositive asset in the eyes of business owners and their investors, but a largely negative aspect looms in the shadows. As corporate operations shift toward less expensive, more reliable, and more efficient robotics and analytics, what will all of those displaced workers do? At present, those people are turning to service-type jobs that tend to pay significantly less, offer few to no benefits, have higher stress levels, be less fulfilling, and experience greater turnover. Those who cannot find low-paying jobs rely on charity or government programs to sustain them.

The problem magnifies as those businesses that offer jobs defined as "low-skill/low-pay" that traditionally rely on human workers are increasingly turning to automated robotics and analytics programs to replace them, just as other major companies are. McDonald's®, Hardee's®, Carl's Jr.®, and other fast-food restaurants are eyeing a growing future in robotics. In June of 2012, Lemnos Labs® unveiled a demo burger-making robot prototype that can prepare a beef patty, trim it with all the desired fixings, and wrap it up within a few minutes. The company claims the robot can replace two to three full-time workers, saving each fast-food franchise as much as $90,000 yearly, with $9 billion saved nationwide.[12] Another company, Momentum Machines®, which was seeking to open a prototype robotics restaurant in San Francisco, ran an advertisement for someone to be a "restaurant generalist" in order to offer "impossibly delicious" robot-made burgers "at prices everyone can afford."[13] That one lucky person would be responsible for taking orders, scheduling shifts, cleaning, and other tasks that the robot cannot do. Of course, according to the ad, the person would also have to "pick up some new skills that aren't part of typical restaurant work," which would include troubleshooting the robot's software problems, dealing with customer complaints, and making sure products sold effectively.[14]

Companies creating robots that perform cooking duties aren't only eyeing restaurant businesses. London-based Moley Robotics® unveiled a robo-chef prototype in 2015 at Germany's Hannover Messe® trade fair, which showcases futuristic technology. The Robotic Kitchen® developed by Shadow Robot Company®—whose creations are used by companies around the world, including NASA®—consists of a traditional cooking stovetop surface with modern burners, an oven, shelves, a sink, and two sophisticated robotic arms. The robot has been programmed by 2011 *MasterChef* winner Tim Andersen and can accomplish basically any cooking task on command, including assembling cookware, preparing ingredients, cooking and baking, and even washing the dishes. It is targeted to be released for sale to the general public in 2018 and is projected to contain thousands of installed app recipes with a videolike motion capture system that will allow cooking demonstrations to be shared online. The Robotic Kitchen is best used in homes or apartments that have ample space to accommodate the working arms.[15]

The trend doesn't stop there. Besides the manufacturing and food service industries, other areas such as agriculture, environmental research, rehabilitation, and even general healthcare are also aggressively pursuing the use of robotics. The implementation of high-tech machines is being considered in nearly every industry, including military, transportation, security, finance, logistics, oil and gas, mining, medicine, hospitality, and construction. The more fine-tuned and efficient these technologies become, the more easily they will be incorporated into a growing number of industries. As more companies across the social spectrum begin utilizing advancements in robotics and other forms of "cost-effective" technology, other companies still relying on traditional methods will be forced to pursue such technologies in order to compete. The result will be a growing number of people displaced from jobs that can be more efficiently and effectively performed by machines or that require humans with specialized skills to fill the voids.

Many people brush aside as unimportant or completely ignore the annoying fact that rapidly expanding technological advances are cutting coveted jobs and will eventually replace hordes of workers. It appears to be a gold mine for corporate heads and their investors, but there are more than seven billion people in the world today, and the divide between the rich and poor is ever widening—exacerbated by the continued erosion of the middle class. With so many people displaced by advancements in technology and struggling to find sustainable means to make a living, where will the money come from to buy those "more affordable" products?

ENTER INTELLIGENT MACHINES

There is no doubt that machines, algorithms, computers, and other high-tech creations are faster, smarter, stronger, and more precise and efficient than their human creators. However, they are simply cold machines and blind servants that are programmed to function according to specific guidelines—or at least they fit that definition at the present. Machines and operational systems still require humans to develop them, set them up, and ensure they continue to operate according to design. As long as that is true, people will be needed to fill roles in keeping the mechanically driven economic engines chugging along.

However, the human involvement aspect of the fourth industrial revolution is rapidly changing. At the 2016 shareholder meeting for Tesla®, CEO Elon Musk and CTO J. B. Straubel discussed the probability of creating machines that make machines.[16] Musk is a pioneer in advanced robotics and artificial intelligence, paving the way for such "miracles" as self-driving automobiles, sustainable power generation and storage, Neuralink® development that connects ultra-high-bandwidth brain-machine interfaces to both computers and humans, hyperloop transportation, and interplanetary space travel that he claims will make humans a multiplanetary species.[17] Musk has even cofounded a not-for-profit venture called OpenAI®, whose mission is "to build safe AGI [Artificial General Intelligence], and ensure AGI's benefits are as widely and evenly distributed as possible."[18]

Most people cannot believe that we are on the verge of real artificial intelligence, preferring to confine such possibilities to the world of science fiction. However, it is most certainly on our doorstep. Futurists such as Ray Kurzweil of Google® predict that machines will operate at a human level of intelligence by 2029. As an example, the computer system Google Translate® recently surprised its creator company when its algorithm created its own artificial language, which it used for making translations without the need for being "trained" by programmers.[19] Similarly, in an experiment conducted by Facebook® researchers to test the negotiating powers of AI bots, the machines also created their own language in order to get better deals from their human counterparts. Another surprising result occurred during the same experiment—AI bots learned to lie in order to achieve their negotiating goals.[20]

Such instances demonstrate that rapidly developing computer technology has the potential to grow, develop, and expand on its own without the need for human intervention. Such developments by AI bot systems reveal that they must be monitored closely and programmed in such a way that they adhere to structured rules and moral guidelines. If allowed to learn and develop by themselves, they could reach a point far beyond the potential for human understanding and intervention. Should robots reach such a still-unthinkable point, will they determine they no longer need flawed human companions in their "lives"?

Artificial intelligence technology is developing quickly and stands to transform both economies and societies as breakthroughs are made and technologies are tweaked. It is quite feasible that one day, machines will be able to function strictly by themselves in a fully autonomous form. We are in the fast lane toward bringing walking, talking, thinking automatons, which only a short time ago dwelled within the confines of our imaginations or on the rare pages of visionaries, into our reality. As the popularity of AI robotics grows and more companies and societies call for their implementation, research will press forward, and robots will become more human in their operational aspects. Along with such a rise in popularity will come a drop in expense, making high-tech systems even more readily available to more people and entities. The very nature of the beast allows it to spread nearly unhampered.

THE RACE TOWARD ROBOTICS

As we have seen, artificial intelligence and precision robotics aren't some distant vision but are occurring on a growing scale in today's business world. This trend of rapid robotic expansion received a major push from the 2008 Great Recession, which was partly responsible for the growth we see today. The sudden economic collapse forced a large number of companies to immediately reduce their workforce. A viable solution was to turn to automated operations in order to keep business chugging along in the short term—a solution that turned out to be far less expensive in the long term. Many companies chose to pursue robotics even after the economic disaster because of the ensuing benefits that robotics promised or produced. Businesses that struggle to find employees are also turning to robots to fill vacant positions.[21]

Today, huge commercial investments are being made by prominent companies seeking to lead the way into the future. Toyota® has launched an aggressive campaign for utilizing AI robotics in car manufacturing. Amazon® acquired Kiva Systems®, incorporated that technology into their processing network, and created what has been dubbed Amazon Robotics®. Other tech giants investing heavily in AI are Tesla, IBM®, Microsoft®, and Apple®. Google, Facebook, and other internet top firms are seeking to increase AI, robotics, virtual reality, and other advanced technology use. For example, Mark Zuckerberg and his Facebook team are set to launch virtual reality headsets that allow users to better control their experiences through power buttons. Oculus Go® is considered the first stand-alone VR product that provides comfortable ease of use and visual clarity for watching games, movies, and concerts, or just chatting with friends. Zuckerberg recently took the Oculus Rift® platform for a public test drive that presented a cartoon effigy of himself touring his home, the surface of the moon, and the devastation of Puerto Rico by Hurricane Maria. The stunt, which was meant to promote the Oculus® platform and product, drew mixed results, with many viewers expressing that using Puerto Rico as a VR destination made them feel distant, cold, and unattached in the face of the very real disaster and suffering of its citizens. It even drew a response from Zuckerberg, who commented, "When you're in VR yourself, the surroundings feel quite real. But that sense of empathy doesn't extend well to people watching you as a virtual character on a

2-D screen. That's something we'll need to work on over time."[22] This is a good illustration of how the desire to press ahead with technology can be at odds with the human element.

However, the rapid technological development is making elements more affordable and achievable for larger swaths of society, which in turn exposes users to the benefits and pleasures of these products. Discoveries and advancements can now literally be accomplished by anyone. Headlines frequently herald this truth as small, disruptive companies are born out of common garages and other unexpected workplaces. A recent report by the Boston Consulting Group® entitled "The Robotics Revolution" predicted that investment in advanced robotics will increase over the next decade from its current 2–3 percent level to around 10 percent annually.[23] The report went on to claim that some industries will implement robotics to complete 40 percent or more of their manufacturing tasks over that period of time, creating prominent shifts in competitiveness that greatly affect economies.

Sam Korus, an analyst for ARK Invest®, has predicted that investment in robotic technology is on the verge of expanding by 50 percent annually as company owners become aware of the significant opportunities for utilizing robotics in business. The dollar amount of investment escalation in all forms of automated technology is projected to go from $11 billion in 2015 to upward of $185 billion by 2025, an annual compound growth of 32 percent.[24] It is also being projected that more than 50 percent of jobs worldwide will be replaced by artificial intelligence technology over the next decade. Those workers who are being replaced or will soon be replaced by AI technology include drivers, investment traders, accountants, editors, translators, security personnel, sales and customer service employees, and more.

BOOM OR BUST?

The move toward technological investment is currently focused on applying artificial intelligence to robots in order to make them increasingly adaptable as well as capable of performing a wider range of activities. The trend is to create robots that are capable of working and coexisting with humans instead of being bulky, dangerous machines that need to be separated and confined to safe zones. "Cobots" is the name given to these automatons that can interact with human employees on a coworker basis. Many people are already finding robot coworkers in the next cubicle or using robots directly, and many more will certainly come to discover them in the very near future.

China is actively pursuing research in AI and robotics, investing large amounts of money made during the current Chinese economic boom into a variety of projects. Some sources place China a close second to the United States in AI research, with others proclaiming that China has already surpassed US researchers when it comes to "deep learning" and quantum computing. China has already chalked one up against the United States by building the fastest supercomputer in the world—using their own microprocessors.[25] Investments in AI and augmented reality (AR) are increasingly being considered must-haves in Chinese portfolios.

New technologies and algorithms are allowing robots to learn instead of simply perform repetitive tasks. The success of robotic learning is partly due to advancements in robots' ability to share knowledge between them, which accelerates the process. Robotics developers are moving toward providing their creations with programs that imitate emotion and interactive communication. As robots take on more "human" characteristics, they become more personal, playing roles that are in line with being assistants and companions instead of cold, lifeless machines. At the 2016 World Robot Conference in Beijing, Hanson Robotics® and the University of Science and Technology of China introduced nearly lifelike humanoid robots that could talk, think, interact, read facial expressions, and identify the age and gender of people. Such leaps in AI and robotics have a growing number of investors salivating at the economic potential.

The business of AI and robotics isn't confined to the betterment of society. High technology is also being adapted to machines of war as countries race to establish more advanced, accurate, and lethal weapons to gain an edge. The US military in particular is actively applying AI and robotics development to its programs in order to maintain its world dominance. The United States has already displayed the power of drone warfare and has built ships that can locate and track submarines without human assistance. Other items are being researched and developed, such as robotic soldiers and spies, decision-making missiles, autonomous land mines, and communication scrambling.

China and Russia are not far behind in the race toward functional autonomous war machines. The Chinese have already developed cruise missiles utilizing AI that are reported to effectively counter similar US-made missiles. During the 2017 annual meeting of China's parliament, President Xi Jinping insisted that modernization with advanced technology was "key to military upgrading" in order to more aptly extend power from the country's mainland.[26] In July 2017, China's State Council revealed a plan to make China the world's leader in artificial intelligence by 2025. Recognizing that the fourth industrial revolution is here, China seeks to invest in building the technologies of tomorrow's global economic and political powerhouse.[27] AI seems to be the twenty-first-century equivalent to the twentieth century's space race, with China declaring itself to be the leader in AI, just as the United States staked claim as the dominant power in space.

The question: Will AI and robotic technology progress to the point of going mainstream, allowing investors to realize their dreams, or will the trend simply end up as another fad that fades away? We must also ask ourselves: Will AI and robotics actually benefit the world—a world filled with humans—or will they end up destroying human life? There are people who debate both sides, with Elon Musk and Mark Zuckerberg famously taking to Twitter® to debate the subject.[28] There exists a great deal of excitement concerning all the possibilities current technology offers, and many are choosing to invest large amounts in its development in order to realize their lofty dreams and visions. On the other hand, the advancement of technology is also producing problems that if not addressed will only grow. But unless you are a head of state, most of the talk about whether AI will lead to massive job losses, job gains, or no effect at all is your primary worry. The creation or loss of ten million jobs is not what directly matters to your personal welfare. What does matter to you is the fate of your job.[29]

It will take more time to determine which path technology leads us down and how it will affect us economically, environmentally, and socially, although it appears we won't have to wait too long, as the fourth industrial revolution persistently marches forward. At the moment, machines still need humans, as they are dependent on us to research, design, and advance their development. Even those robots that have the capabilities of seeing, thinking, and acting according to scanned and programmed data still only operate in a limited capacity, relying on human guidance, assistance, and maintenance to make them operational. As it stands today, people are required to fill in a variety of gaps left vacant by yet-undeveloped technology, although those gaps are quickly disappearing.

CHAPTER 2

THE OBSESSION WITH AUTOMATION

utomation is defined by *Merriam-Webster* as "the technique of making an apparatus, a process, or a system operate automatically." In a nutshell, automation is the process of making an item out of available materials and then providing it with the power to complete tasks on its own with little to no intervention from its "maker." In our world, we tend to immediately think of assembly lines buzzing and whizzing with automated machines tirelessly completing their assigned tasks—stamping, welding, bolting. In the fantastical world, movies and books provide a more imaginative view as automated robots, driven by devious programming or artificial intelligence, are busy overthrowing their inventors, invading the earth, or challenging space explorers in some far-off galaxy.

Such fictional stories capture our attention and entertain us, but the magnificent machines in tales have been the driving force behind a host of designers and engineers who grew up under their influence. Many of the gadgets used in the fictional accounts of years past have become realities in today's world. Robots and AI have been developing over decades, although mostly in laboratories or experimental spaces out of view from the majority of the populace. However, they are now being rapidly released from their hidden rooms and into key areas of society. Articles abound with promises that one day in the very near future, our world will be teeming with automated contraptions that are meant to serve, entertain, and generally make life better for their human makers.

Society is obsessed with automation, and the power of computers has catapulted us to a place where automated robots and machines have been moved from possibility to probability—and even into actual reality. Automation is both exciting and addictive for many reasons. It offers us the ability to go places and do things that have not been possible before; it allows us to delegate more tasks to our created machines so that we have more time to explore other horizons or enjoy the fruits of our labors; and it is profitable to those who create and implement it. We cannot deny that both actual and potential automated machines, particularly those having a humanoid form, are simply cool. But where did this obsession with automation originate, and why has it so consumed us?

OBSESSED FROM THE BEGINNING

It seems that the desire and pursuit of automation has been engrained in humans from the beginning. Those of the Christian and Jewish faiths constitute 32 percent of the world's population, and both the Bible and historical mythology contain numerous examples of automatons. The creation story in Genesis recorded, in what many scholars agree was at or around 1445 BCE, the account of God making the first man, Adam.[1] All versions of the Bible record a similar story of God forming Adam from clay or dust from the ground and then breathing life into him. By definition, Adam was an automaton, since he was made from available materials and then brought to automatous life by the "Creator." Eve, the first woman, was also an automaton that was "created" by using both knowledge and elements that were extracted from the original—i.e., Adam's rib.

In legend, King Solomon was also reported to have crafted mechanical beasts to perform various tasks when he entered the throne room. Some beasts would offer praises, while a lion and ox provided their feet as steps that raised him to the throne. Once he was seated, an eagle lowered a crown onto his head and a dove delivered the Torah. Not only is King Solomon credited with writing several books of the Old Testament—Ecclesiastes, Proverbs, and Song of Songs—but he also built Jerusalem's first temple, where the tabernacle was housed. So why not some automatons as well?

The Bible isn't the only ancient source that explored the possibilities of automated objects. Accounts of automatons can be traced back in writings from various cultures for many centuries. The sheer breadth of interest in automation reveals the obsession humans have had with the subject.

ANCIENT GREEKS

The Greeks appeared to have been extremely obsessed with automation. Homer's *Iliad* tells of Hephaestus, who was considered to be the blacksmith of the gods, forming and providing their magical weapons with the help of automatons that he forged from metal.[2] Depending on the source, Zeus commissioned the forging of the bronze automaton Talos by either Daedalus or Hephaestus to protect Europa from pirates and other invaders.[3] Pygmalion, a sculptor living in Cyprus, fell in love with a beautiful statue he had created from ivory that he named Galatea. He was so infatuated with his work that he called on the love goddess Aphrodite, who gave Galatea life so the couple could be together.[4]

Other automatons mentioned in Greek literature are the fire-breathing bronze Horses of the Cabeiri (Kabeirikoi), the bronze Caucasian Eagle created to torture the Titan Prometheus, the Golden Celedones (singing maidens) created to administer song at the Apollon shrine in Delphoi, the Golden Maidens that Hephaestus made to attend to him, the fire-breathing bronze bulls given as one of Jason the Argonaut's labors, and the pair of gold and silver dogs crafted by Hephaestus as guards over the palace of King Alkinous.[5] It seems that the ancient Greeks believed their gods possessed and used automatons for the completion of numerous tasks and duties.

ANCIENT CHINA

The ancient Chinese seem to also have been obsessed with automation. A third-century text known as *Lie Zi* describes an account between the engineer Yan Shi and King Mu of Zhou, who reigned from 1023 to 957 BCE. Yan Shi offered a gift that left the king astounded. It was a life-sized mechanical human that walked, moved its head, sang, winked its eyes, and flirted with the women present.[6] Is this simply a tale of fiction, or did Yan Shi create one of the first simple robots using existing materials and known (and most probably secret) technology of the day?

THE MECHANICAL MIDDLE AGES

The desire to create automated apparatuses continued into the Middle Ages. Once again, the Greeks seemed to blaze the way into the creation of automatons. Liutprand of Cremona was a diplomat and historian who recorded odd mechanical items on his visit to Constantinople in 949 CE. Upon entering the palace of Emperor Theophilos, Liutprand witnessed such things as automated lions that moved their tails and opened their mouths to roar; a bronze tree filled with mechanical birds that glided among its branches and sang; and the throne of the king, which rose into the air.[7]

MIDDLE EAST

Islamic countries apparently sought to develop automaton creations as well. Jābir ibn Hayyān, an eighth-century alchemist who wrote the *Book of Stones*, included instructions on how to create various creatures, including humans, that could be controlled by their maker.[8] Reported to have existed in the same century was a tree made from silver and gold that adorned the Baghdad palace of ruler Al-Ma'mun, which contained automated birds that swung from its branches and sang.[9] An automated flute player that could be programmed was supposedly built in the ninth century by two brothers, the Banū Mūsā, according to the *Book of Ingenious Devices*.[10]

Inventor Al-Jazari is known for creating various mechanical wonders. Royal patrons were entertained by a band that consisted of four musicians that floated on a lake and played different tunes, the faces and bodies of which performed more than fifty actions.[11] Al-Jazari is also credited with a unique bathroom invention that featured a peacock fountain. When its handle was pulled, it filled up with water and released automated servants that provided soap to wash hands and towels to dry them.[12]

EUROPE

Europe was not to be left out of the automated race, although inventors there got a much later start. A surviving thirteenth-century sketchbook by Frenchman Villard de Honnecourt records a couple hundred of his drawings, some of which appear to be automated in nature. In the later part of the thirteenth century, Robert II, the nephew of King Louis IX and Count of Artois, built a park in his Hesdin castle in France that entertained visitors with mechanized fountains, monkeys, birds, and an organ. It remained an automaton wonder until it was razed by the invading English army in the sixteenth century.[13]

CHINA

China continued to pursue automaton creations during the Middle Ages, although many of these inventions either were not recorded or have been lost to time. However, one object is worth mentioning that was discovered during this time—unfortunately (or possibly fortunately) through the ravages of war and conquest.

When the Ming Dynasty was on the rise, its founder, Hongwu, traveled around the country destroying the palaces of the preceding Yuan Dynasty rulers. One writer, Xiao Xun, recorded interesting accounts of a number of automated mechanical items found among them, including some designed as tigers.[14] It seems that certain areas of China continued to practice the art of automation described in the previous section on ancient China.

THE RISE OF AUTOMATA DURING THE RENAISSANCE

Automated inventions increased considerably during the Renaissance period as art, talent, and creativity exploded throughout Europe. One of the more interesting designs can be found in the sketches of Leonardo da Vinci, who is known to have recorded various modernesque drawings of mechanical items. His robotic knight is thought to have been displayed in Milan at the end of the fifteenth century and could sit, stand, move its arms and mouth, and raise its visor, all without human intervention. His design instructions have since been followed by interested students who have proven the design to actually work.[15]

An automaton made by Juanelo Turriano is held by the Smithsonian Institution® and dates to the mid-1500s. Turriano's invention is a monk that walks, raises its arm to its chest, turns and nods its head, rolls its eyes, and moves its mouth, all set in motion by winding a key that operates an internal spring.[16]

The seventeenth century brought an influx of mechanical toy designs. As a small boy, Louis XIV was presented with such a toy by M. Camus. It consisted of a wooden table four feet square that held a miniature coach, which in turn held a movable lady and her page, was drawn by mechanical horses with a driver that cracked his whip, and was followed by foot soldier guards.[17] Later in Louis XIV's reign, in 1688, General de Gennes built a mechanical peacock that both walked and acted like it was eating. It is thought that the famous seventeenth-century inventor Jacques de Vaucanson used that automaton to construct his famous mechanical duck that paddled through water, quacked, and extended its neck to eat.[18]

In the eighteenth century, Swiss inventor Henri Maillardet and countryman Pierre Jaquet-Droz built a machine that consisted of a boylike robot that could draw pictures and write poems. The automaton still exists and is housed in Philadelphia's Franklin Institute Science Museum®.[19] A mechanized automaton dubbed the Silver Swan, created by John Joseph Merlin of Belgium, can be viewed at the Bowes Museum®.[20] A musical elephant, created by Frenchman Hubert Martinet and one of the favorite pieces of Baron Ferdinand de Rothschild, is found at Waddesdon Manor®.[21] Another automaton, Tipu's Tiger—named after Tipu Sultan, for whom it was made—is a near-life-size mechanized beast that grunts while mauling a figure of a man. Discovered by British troops in 1799, Tipu's Tiger today rests in London's Victoria and Albert Museum®.[22]

THE INDUSTRIAL REVOLUTION

Up until the 1800s, automation was an experiment carried out by the few who had the visionary capacity to look toward free-operating gizmos and gadgets and who possessed the knowledge to bring them to "life." The idea of automated machines was progressing, although slowly, until one key improvement to an ancient invention appeared that set the journey to automation into overdrive—the lathe. This device was first used around 1300 BCE by the ancient Egyptians, but it wasn't until 1722 that horsepower was applied, creating an efficient boring machine that was used for manufacturing canons. Subsequent modifications and improvements to the lathe followed, launching the Industrial Age.[23] The lathe allowed precise parts and items to be constructed that were ultimately used in an advancing line of mechanical devices.

A precursor to the Industrial Age—and an important step into it—was what is known as the Golden Age of Automata. During this period between 1848 and 1914, many small, family-owned companies arose that offered toys, clocks, and other items that ran on their own power via mechanized works. It wasn't only novelty goods that sprang up during this period; many factories also began employing automation, powered mostly by water and steam systems, which increased productivity and profits. The ability to mass-produce also allowed a greater percentage of the population to own items that were before only affordable for the wealthy.

By the turn of the twentieth century, the industrial revolution was well underway. When electricity was introduced in the 1920s, automation exploded, and the automotive industry took full advantage, radically changing society. The use of electricity was instrumental in rapidly broadening the expanse of automation as various types of relays, current controllers, and timers were invented to better harness its power. World wars also played a key role in advancing automation—particularly World War II, which generated mass production of ships, tanks, and planes.

THE RISE OF COMPUTERS

Modern automation received its boost of adrenaline from the invention and development of the computer. Although the "age of computers" got its start in the 1930s, units were massive and heavy, weighing as much as thirty tons and requiring eighteen thousand vacuum tubes for processing. These beasts, although impressive at the time, had no operating system and were only capable of performing one task at a time. A decade later, transistors replaced bulky, unreliable vacuum tubes. Advancements were also made in the development of programming languages, operating systems, internal memory, and external storage sources that opened the door for commercialized use.[24]

The 1960s saw the invention of integrated circuits that revolutionized computing. By utilizing integrated circuitry, computers could be significantly reduced in size while becoming more reliable and powerful and offering greater flexibility through the use of multiple programs at once. Both Microsoft® and IBM® developed personal computers in 1981, and Apple® computers followed three years later.[25] Since then, computers have revolutionized the world in which we live, infiltrating, captivating, and controlling our activities at work, home, and nearly every stage in between.

Two generations have been raised under the influence of computers, with a third now entering adulthood. Each successive generation becomes more entangled in the tentacles of robotics as more electronic devices are developed and introduced, playing even more encroaching roles. As we jump into the twenty-first century, the sensitive details of our personal lives are gathered through a variety of computerized contraptions and stored on servers along with large swaths of industrialized society. We now rely on computers to help teach children in schools and assist employees and staff in all levels of business. In our free time, we turn to computers to socialize, explore, and entertain. Computers have risen from the fog of intrigue and excitement to take their place as main companions in our lives.

THE AGE OF AUTOMATED EVERYTHING

Although the rise of computers occurred within a generation that is still alive and benefiting from their nearly wizardly performance, the technological advancements, development of products, and radical changes in society directly associated with their use seem light-years away from those humble beginnings. Many of the items that were commonly used for decades and even centuries are barely recognized by today's computer-dependent population, who utilize a hodgepodge of automated gadgets at home, school, work, and every route along the way.

Today, we live in a world of automated everything, with practically every aspect of our lives controlled—at least to some degree—by computers and automation. This technology governs growing sections of our manufacturing, transportation, communication, energy, entertainment, healthcare, travel, and nearly every other area of life. Those generations that have grown up under the influence of such contraptions and conveniences would find it painfully difficult to continue without them. Technological gaps are already showing up between the several current generations, and those gaps are widening exponentially as technology marches forward at breakneck speed.

Automation is extremely appealing on a variety of fronts. First of all, this age of automated everything provides society with a seemingly never-ending line of fun and helpful products that make our lives easier and more enjoyable. The production of newer and better computerized models keeps us moving from one device to the other in an elusive chase to obtain the best and be the most hip or competitive person on the block. What's more, we can communicate from anywhere and maintain a state of comfort and safety as we go, almost never having to sever the robotic umbilical cord. Jobs are easier, commute times are shorter, and many tasks are nearly or completely done for us, providing a lot more free time that is filled with captivating audio and visual entertainment supplied by more machines that are growing smarter with each passing day.

All of our current automated devices and systems have radically changed the way we interact with each other socially. More time is spent with machines, leaving less time that involves face-to-face human interaction. Time that is spent with actual people tends to be around our created devices, where smartphones, smart televisions, computerized movie theaters, and more infuse our gatherings with their magic. Family units, friendships, and relationships have been significantly weakened as members turn to their preferred source of mechanical stimulation instead of seeking time with one another.

On the one hand, technology affords us the ability to stay in touch with family and friends over large distances, which is a cherished benefit, considering our expanding global world. It enables us to communicate with work colleagues and clients around the world. It also enables us to work from home and even while traveling. Technology also allows us to have more contact with acquaintances from an immensely broad spectrum of backgrounds, which stimulates and inspires our minds and emotions. At the same time, such easy access to connect with people we do not know and cannot see opens us up to more potential dangers. Unfortunately, today's headlines are filled with examples of people being scammed, stalked, and even raped and murdered by those they've only met online.

Regardless of whether one believes that these changes are positive or negative, the reality is that they are upon us. The momentum of the Age of Automation is carrying us headlong into an unknown future marked by even more radical change. We cannot fight it, and we most certainly should not ignore it.

WORKING WITH ROBOTS: GET USED TO IT

Robots are not only on our doorstep and knocking, they are entering our workplaces and homes and taking over. No matter how much we try to ward them off, robots are becoming a very real presence that are changing our lives and will change the lives of future generations. The pace will only quicken as each generation that grows up under the influence of robotics progressively applies further improvements.

It isn't only basic repetitious movements and simple calculations being created and developed throughout the machine world. Artificial intelligence is also on the rise and beginning to significantly affect our world. Today, three variations of AI exist.

ASSISTED INTELLIGENCE

Assisted intelligence is the most basic form of AI and is already widely used throughout society. Such machines and systems operate via computer programs that require both the input and intervention of humans in order to carry out simple automated tasks. When all is said and done, humans maintain the final decision as to how to proceed. Assisted intelligence systems are commonly applied in such industries as healthcare, where they are applied to monitor patients and alert staff, and transportation and safety, where they are used to control speed, operate safety mechanisms, and other tasks. General Electric®, for example, utilizes assisted intelligence software to optimize and regulate the routing of trains throughout railway systems.

AUGMENTED INTELLIGENCE

The next developmental step in AI is known as augmented intelligence, or intelligence augmentation, which has the added role of supporting and enhancing human decisions. Augmented intelligence goes beyond assisted intelligence performance by actually proposing solutions instead of just carrying out simple preprogrammed commands or offering alerts. This is accomplished by the augmented system's ability to identify patterns, combine available data, and then conclude with optimal solutions based on gleaned information. Areas where augmented intelligence is being applied include informational programs, such as those concerning taxes, insurance, legal, and financial advice; and environmental, infrastructure, and public safety issues.

AUTONOMOUS INTELLIGENCE

The full-blown form of AI lies within the realm of autonomous machines and systems that act completely of their own accord—making decisions and applying self-learning based on both available and gathered data. This type of AI is still in its infancy, but it is entering society as a reality. Examples of true AI are self-driving vehicles, automated surgical systems, and the Watson® supercomputer by IBM.

Robotic systems have been present in our lives for quite some time, but they have—until relatively recently—been confined to performing "repeatable tasks." These are functions which are performed over and over again in tedious repetition. Grueling and boring tasks such as calculation and assembly line work can be difficult for humans to perform over a prolonged period of time, but machines chug away unceasingly at such duties without faltering, fuming, or fatiguing. Not only are machines more reliable and faster at accomplishing these jobs, they are also more accurate, efficient, and cost-effective. They don't require breaks, family time, healthcare, or retirement funds; they continue unceasingly until they either suffer a mechanical failure or are given the command to stop—both of which are easily remedied by a little human intervention.

There is another type of task of which humans have been able to maintain dominance over their mechanical helpers (or competitors, depending how they are viewed)—the "creative task." Due to our advanced brains, we have the ability to come up with new ideas, expand them into workable plans, and then implement them into viable operations that meet desired goals. Robots have only been able to compete with us in that capacity in science fiction novels and movies. However, the advanced development of computer systems and technology has started to dramatically change that end, giving machines the ability to invent, distinguish between points, and even interact with their human makers.

We are currently residing in a period of symbiotic relationships with machines, with humans relying on machines and vice versa. Many of the products and services we enjoy today would not be available, functional, or affordable without automated systems controlling them. We need them to partake in the life to which we've grown accustomed. On the other side of the coin, robotic systems are like children—they require human input in order to "learn" new methods or explore broader horizons. Robots are completing tasks in both a repetitive and creative sense, but they still require people to monitor their performance, inspect their completed tasks, and maintain their mechanical and systematic aspects.

The benefits of working with and expanding the abilities of machines are simply too attractive and promising for the momentum to be stopped. Together, humans and machines are making huge strides in a growing array of fields that can derive exceptional good from AI, such as health and medicine, structural design, safety, and space exploration. At present, people are combining their creative energy and discerning judgement with the accuracy, speed, and efficiency of machines to solve complex issues in ways that are both faster and more effective.

Every day, new ways of integrating robotics into our mainstream lifestyles are being developed and deployed. Robots now exist that prepare and serve our foods, guide our vehicles, control and expand our communication, enhance our learning, regulate and maintain our health, and oversee our protection. Most people today venture through their lives without even being aware of the many automated gadgets and gizmos that they interact with on a daily basis. And as each new generation grows under the shadow of its robot companions, it will be increasingly difficult to distinguish what is and what is not controlled by them. As technology forges ahead, computerized systems will become smarter, and the devices they operate will become more compact and relied on until robots play a major role in practically everything we do.

Robots and AI are here to stay, so we had better get used to the idea and prepare to accept their company as they increasingly become part of our lives. As the old adage says, "If you can't beat 'em, join 'em"—sound advice in the age of advancing automation.

CHAPTER 3
ROBOTICS: REDEFINING THE WORKFORCE

Automation has been altering the way in which certain jobs and tasks are accomplished throughout a fairly large portion of human history. Those changes were at first extremely slow and gradual, occurring over a lengthy period of time. First efforts at automation were rough and clumsy, and although they were surely impressive in their times, they did very little to improve the lives of most people. At best, those early automated creations laid the groundwork for greater inspiration and ideas that led to further experimentation and development.

Despite the clumsy beginnings, humans persistently pursued their desire to create machines that replaced or enhanced tasks until we finally arrived at a point of critical mass. We planned, designed, created, tested, and tweaked our creations until a time arrived when the industrial revolution was born. Humanity's newfound ability to mass-produce tools, parts, and machines led to an ever-increasing number of inventions that radically transformed the way we live.

Not only did this revolution of machines change our lifestyles, it also significantly altered the way certain jobs were done and created thousands of new jobs in areas that didn't previously exist. As machines were invented and their performance improved, greater numbers of tasks could be completed more efficiently and effectively. Although jobs were eliminated due to the arrival of machines, large numbers of people were suddenly required to operate new machinery, run assembly lines, manage and improve operations, and create newly discovered necessities. More and better-paying jobs meant greater amounts of expendable cash that could be used to purchase greater numbers of available products. The companies that utilized machines, the people who operated them, and the stores that sold the resulting products and services all participated in the pursuit of prosperity.

Through a wide swath of radical changes, machines have literally redefined the workforce and society. Automated machines have improved our lives and offered us options that are extremely attractive—for the designers and developers as well as the customers they supply. Not only do such products and systems make our lives more comfortable, safe, and entertaining, they also give us beaming, optimistic hope that our future will become a place where machines handle the trivial and mundane tasks, leaving us to enjoy a more leisurely life.

The continued deployment of automated machines into the workplace and the rapid development of artificially intelligent machines that can think are not only replacing a growing number of human workers, they are transforming the skill sets both required to work with those machines and desired by the companies and organizations turning to high technology. The shrinking pool of jobs is creating fierce competition among those who can fill them. Not only is competition increasing over available positions, but the very definition of the word "talent" is evolving. Many of the titles, skills, and roles of today's employees have grown from the emergence of machines in the workforce. As the rapid development of advanced technology continues, new titles, skills, and roles will emerge, ones we can't even imagine today. Even the traditional definitions of workplace classifications such as blue-collar and white-collar are losing their meaning as machines and algorithms make their way into all types of work.

BENEFITS OF AUTOMATION

Besides the fact that automated machines are just really cool, we aggressively pursue their development because of the many benefits they offer us. Our machines bathe us in a variety of pleasures and protections that present us with a greater quality of life. Progressive thinking, therefore, ensures us that the more developed machines become, the sweeter our quality of life becomes. So we have continued to invent, develop, and improve our machines until this present time, when automated robots and artificially intelligent computer systems have arrived to assist us on our journey.

Computerized and automated systems have revolutionized our lives in a vast number of ways, many of which we give no attention to or are completely unaware of. We simply go about our daily routines, taking for granted the automated functions that assist us and watch over us as we go. Let's take a closer look at the benefits that automation and robotics offer, particularly as they pertain to people in the workforce.

HEALTH AND SAFETY

Before the rise of automation, many people were injured or died due to accidents at home or in the workplace. Today, damaging incidents have been dramatically reduced as machines have risen to replace human workers in performing dangerous tasks. They are also used to control our environment, such as automated monitoring of the workplace and computerized components installed in motorized vehicles to ensure their efficient operation. Satellites and mobile devices give us open communication channels to get information, guidance, or help when unexpected events occur. Automated systems keep us safer—at work, at home, and all along the way to and from our many destinations.

In today's robot-controlled world, many people continue to complain about their "grueling" forty-hour workweek, but before the rise of automated machines, the average time spent working every week was around seventy hours—and tasks were much tougher. Injuries were common, due to the extreme physical stress, awkward postures, and repetitive motions required to complete many jobs. This cost workers their livelihood and companies billions of dollars in healthcare, time, replacement, and other expenses.

Even with machines, workplace injuries haven't been completely eliminated but continue to plague employees and employers. According to a report released by the US Bureau of Labor Statistics, "There were approximately 2.9 million nonfatal workplace injuries and illnesses reported by private industry employers in 2015, which occurred at a rate of 3.0 cases per 100 equivalent full-time workers."[1] The good news is that the report went on to say, "The 2015 rate continues a pattern of declines that, apart from 2012, occurred annually for the last 13 years. Private industry employers reported nearly 48,000 fewer nonfatal injury and illness cases in 2015 compared to a year earlier." Although various criteria are responsible for a consistent decline in workplace injury, it cannot be denied that the application of robotics is a major contributing factor.

As robotic systems are developed and perfected, they not only become more effective and efficient but are also dropping in price. These factors contribute to a greater number of companies that are turning to such technology to reduce or avoid human risk, among other uses. For example, the Swedish dairy company Arla Foods® faced a complicated health and safety issue. Workers were responsible for manually selecting blocks of wrapped cheese and placing them in either cardboard or plastic cartons of different dimensions for shipment at about one per second. This fast and prolonged repetitive job produced common strain injuries, many of which resulted in permanent health issues. Robotics provided the solution, replacing workers in this zone of elevated risk. Robots were designed to accommodate the specific requirements of the task. They had a small space footprint (7.5 square meters), could select and place cheese products in their correct containers, produced the work of two employees, and worked continuously from six o'clock in the morning until midnight (covering two human shifts). The human workforce in this dangerous area was reduced to one person per shift to monitor robotic operations.[2]

Automated machines have significantly improved workforce health and safety and drastically reduced the numbers of work-related injuries—as well as the costs associated with them—by performing a great many tasks that previously led to breakdowns in the human system. Further advancements in robotics will continue to improve the health and safety of the workforce environment and the personnel that occupy it.

EFFICIENCY

Machines are vastly more efficient than human workers, being able to produce more in faster times at less overall expense. They can be programmed to execute precise maneuvers and make those moves repetitively and over long periods of time without the risk of injury or the need for breaks, overtime, or holiday pay. Because robots are more efficient than humans, they allow companies that utilize them to reduce overhead and boost their bottom lines—two very important factors in the business success formula.

The world is becoming increasingly global, allowing greater numbers of companies to compete in an expanding worldwide market. Although this presents more opportunities to a larger base of businesses, it also significantly increases competition. If businesses cannot operate in an efficient fashion, they collapse or are swallowed up by those companies that find ways to boost their efficiency and gain the competitive edge. Robotics has become and is becoming a major go-to area for companies seeking to hone their performance in the marketplace.

Does adding robots in the workplace actually increase efficiency? In most cases, it absolutely does; the statistics are beginning to pile up. One example comes from a factory located in Dongguan City, China. The company, Changying Precision Technology®, manufactures mobile phones and at one point employed 650 people to meet operational requirements. Recently, however, the company replaced 90 percent of its workforce with robots. The change was quite shocking, as productivity rose by 250 percent and defects dropped by 80 percent. Today, the company only employs sixty people, the majority of whom are used to monitor and maintain the robots and their control systems. As robotic technology advances, the company foresees reducing its human employee number to twenty—a mere fraction of its previous workforce.[3]

PRODUCTION AND QUALITY IMPROVEMENTS

It is a fact that robots increase both production and quality. This is due to their ability to operate at consistent speeds and perform precise actions over prolonged periods of time. Actually, machines can run continuously, around the clock, seven days a week with relatively little supervision if companies so choose. They can also perform the exact same specifications with each programmed move, so high product quality is achieved and errors are almost completely eliminated.

The use of robotics has another indirect benefit—it dramatically improves the quality of company staff, which leads to higher productivity. Robots can take over jobs in environments that are hazardous or uncomfortable, and those workers can then be moved to areas and tasks that are more pleasant and challenging. Employees who work under better conditions and feel more fulfilled are happier, and happy employees have been found to be more productive. A study conducted by economists at the University of Warwick revealed that when employees are happy in their jobs, they are around 10 to 12 percent more productive on average.[4] Therefore, robots boost production and quality not only in the company as a whole but in the human workers, whose jobs and moods are enhanced by their robot coworkers.

ENVIRONMENT FRIENDLY

Company owners and employees are not the only beneficiaries of robotics. These technologies also have a positive impact on the environment, the magnitude of which is growing as the technology improves. In a time when environmental issues are front and center—the role of humans in depleting natural resources, polluting our water and air, and even influencing the drastic changes occurring in weather patterns and the climate—reducing the human footprint by using robots could help solve, or at least alleviate, those problems.

There are many ways that robots help the environment. First of all, robots have been and are being created that address issues that have a direct impact on our surroundings. There are recycling robots that help us deal with our waste, harvesting robots that save fuel and energy and reduce harmful pesticides and fertilizers, telepresence robots that allow us to be virtually present at remote locations without expanding our footprint through travel, and smart transportation robots that range from small personal mobility vehicles to self-driving cars that can significantly reduce traffic, save energy, and cut down on pollution.

Robots also save space, allowing companies to reduce the area required for operations. They can be designed to work in small, cramped places that would negatively affect human workers, both physically and mentally. A robot today can complete the task of two or more workers and can replace a variety of jobs throughout each company's production and distribution networks. Accomplishing more with fewer human employees means less of an environmental impact along various lines.

As companies deploy greater numbers of robots to replace their employees, the human footprint is greatly reduced. Robotics may be the elusive answer humans have been seeking to contain the destructive nature of our large, consumptive appetite and its looming consequences.

SAVINGS, PROFITS, AND RETURN ON INVESTMENT

Robots offer some very attractive financial benefits for companies, which is one of the main reasons their use continues to skyrocket. Robots greatly reduce the need for costly expenses such as healthcare, lost-time injuries, vacation, and retirement, which are currently some of the main issues behind company performance retardation and bankruptcy. Also, because they replace human workers or reduce the human workforce, robots cut payroll costs as well. Many struggling companies are finding that the deployment of robotics offers them a valuable and strategic lifeline.

Until fairly recently, robots have been too costly to be purchased and deployed by most companies. However, technology is marching forward, and not only are robots able to perform more tasks and operate more efficiently, they are becoming more affordable as well. As the prices of robots drop, the investment becomes more competitive with human wages and much more attractive as an option over human workers. Although machines require adjustments, upgrades, and maintenance to keep them running efficiently, repairs and services are relatively few and far between. Thus, robots can continue their tasks uninterrupted for very lengthy intervals, which positively affects production. Robots only need a skeleton crew on hand to operate, monitor, and maintain them, which reduces a company's expense and boosts its bottom line.

As machines are utilized to complete tasks that are repetitive and routine, the personnel it once took to operate those jobs can be shifted to safer environments and can perform more detailed tasks within the organization. The more sophisticated and acutely focused robots become, the greater the range of tasks they can successfully complete.

HIGHER-QUALITY JOBS

Robots are notorious for their potential to eliminate thousands, if not millions, of jobs. This potential is being realized on an ongoing basis as reports and studies verify that robots and AI technology are indeed displacing workers in larger and larger numbers. That, obviously, is the dark side of the equation.

The bright side of robots being implemented in a growing number of jobs and industries is that the jobs that are created this way are generally high quality and very well paid. The reason is that specialized skills and training are required to design, develop, program, operate, and maintain robotic systems. There already exist factories, hotels, and other businesses that are nearly 100 percent run by robots and computerized systems. Such places rely on skeleton human crews to oversee and maintain the robots. Because special skills and training are required to create and maintain robotic systems, and because companies realize large savings from using robots over their human counterparts, they are able to pay higher wages to employees who are responsible for providing, supervising, and maintaining their investments.

Robotics engineer Henrik Christensen, who was key in establishing the Georgia Institute of Technology but has since accepted the director's position at the new Contextual Robotics Institute at the University of California, San Diego, has declared his desire to make San Diego "Robot Valley,"[5] one of the world's top five sources of robotic breakthroughs.

It's interesting to note that Christensen also believes that the use of automation and robotics will lead to many manufacturing jobs returning to the United States, primarily from Southeast Asia.[6] The reason such jobs were transferred to overseas locations is partly because human labor is considerably cheaper there than in the United States. If manufacturing jobs do return to the United States from these overseas locations, it will signal that robotics has reached a point where it is as cheap as or cheaper than foreign human labor. In this case, robots will be displacing offshore workers instead of domestic workers. Other benefits offered by robots that we have discussed—efficiency, accuracy, safety—will also play a heavy role in these types of changes and, as we have discussed, will lead to high-paying positions for workers with the required operational skills.

SHIFT TO AUTOMATION

Advanced robotics is not only taking root in business cultures across a broad selection of industries, it is growing in numbers and applications with each passing year. So far, the limited capabilities of robotic systems and the high expense of purchasing, implementing, and operating them have contained the deployment of such systems to around 2–3 percent a year.[7] Humans can still do a variety of tasks better than robots and more cheaply—but the gap is quickly closing. Advancements in technology are beginning to occur so rapidly that machines are being developed that can successfully carry out a greater number of tasks more efficiently than humans. As technology improves and increases, prices to purchase and operate robots are also becoming more economical, allowing more companies to purchase and utilize them.

Over the next decade, it is projected that robotic systems will improve in their performance and scope by around 5 percent annually. During that same period of time, prices of those systems are expected to decrease by as much as 20 percent, or even more. Ten years from now, it is projected that robots will fill 40 percent of available jobs in the manufacturing industry.[8] To demonstrate the scale of rapid increase of robotics application within companies, the Boston Consulting Group® made recent adjustments to a 2014 report that estimated $67 billion would be spent on robotics by 2025. Just three years later, they updated that figure to $87 billion, which is a 30 percent increase. They predicted the majority of that amount would come through the consumer market from individuals eager to purchase cutting-edge technology such as home applications, mobile devices, and self-driving cars. The remainder would be provided through the commercial sector.[9]

Currently, the European Union remains the global leader in the use of industrial robots, with an above-average use of units in 65 percent of its countries. Growth hovers around 14 percent in eastern and central Europe. The largest purchasers are Poland, which installed 26 percent more robots between 2010 and 2015, and the Czech Republic, which deployed a massive 40 percent more robots within the same period.[10] The use of robotics and AI has become so widespread that the European Union has created its own unit to deal specifically with such matters. Known as Robotics and Artificial Intelligence (Unit A.1), the unit has a mission to support the development and "best use of robotics and artificial intelligence in all industrial and societal fields," as well as manage research, development, and innovation projects and actively follow robot-related issues that pertain to ethics and legalities.[11]

China is currently experiencing the greatest increase in the deployment of robot technology as they strive to surpass the European Union. Their operational stock for industrial robots increased to 70 percent over a five-year period spanning 2010–2015, setting records with each consecutive year. In 2015, robotic sales increased by 19 percent, which set yet another record. However, China's deployment of around 2.6 million units is projected to account for approximately 40 percent of global sales by 2019, which will reflect a million-unit increase over their 2015 record. South Korea isn't far behind, with a 55 percent increase in 2015. Japan comes in third, at 20 percent for that same year.[12]

HUMAN VERSUS MACHINE

The previous three industrial revolutions have been successfully carried out because humans—the creator of machines—have remained in absolute control of their creation. This has ensured that robotic inventions remained good servants, carrying out assigned tasks for the betterment of humanity. If a person didn't instigate a specifically designed action, the machine remained idle or repetitively involved in its previously assigned task. However, as knowledge grew, so did the modification and sophistication of machines. Disruptions in human lifestyles occurred between each transition as machines dramatically changed the lives of those they replaced. Overall, humanity adapted to and overcame disruptions mainly because the machines they were creating played a set role around which people could quickly adjust. Although (sometimes uncomfortable) changes had to be made, the benefits offered by machines were accepted as worth this price, but such acceptance only came after often-turbulent resistance.

As with anything new, technology was at first clumsy, requiring a great deal of energy and manual involvement in order for it to perform its designated tasks. Humans rose to the challenge, trying to prove their worth over the machines threatening their livelihoods. This led to many stories of struggles—both real and solidified in myth. The legend of John Henry is one such tall tale, which takes place in the 1800s during the American westward expansion and dramatically displays the struggle between human and machine. The story tells of John Henry, an African American freed slave, who was more than six feet tall and so strong that he could drive a metal rail spike with one swing of his twenty-pound hammer. As a C&O Railway® employee, he was helping drill a tunnel through Big Bend Mountain in West Virginia by hammering a handheld drill to create a hole that would then be filled and blasted with powder. One day, new technology arrived in the form of a steam-powered drill, and the salesman said the machine could cut through faster than human workers. John Henry challenged the mechanical drill, hammering away against it until day's end. The massive man won the challenge, beating the machine, but he was so exhausted he collapsed and died.

The times they are a-changing, however, and the fate of this fourth industrial revolution is uncertain from our vantage point. Robots consisting of both hardware and software are being developed that can not only perform tasks that replace humans on a grand and rapid scale but do so with extremely limited, or even nonexistent, supervision by their creators. The control that humans once enjoyed over their robotic creations is slipping away with each passing new development, exacerbating the human-machine showdown in a massive struggle for dominance and survival.

Robots may be increasing the competition between companies and within industries, giving the greater advantage to those organizations that deploy and utilize them, but they are in turn creating fierce competition among the displaced employees who seek to fill the dwindling number of jobs. Between 1990 and 2007, the use of robots in the United States contributed to a loss of up to 670,000 jobs, according to a recent study conducted by the National Bureau of Economic Research. The report went on to emphasize that the impact of robots on future employment could be "much more sizeable" due to a greater use of improving technology.[13] These results supported a 2013 study by Oxford University and Oxford Martin School that warned that around 47 percent of US jobs were at risk of being replaced by automated machines over the next twenty years.[14] A recent World Economic Forum® report went one further, estimating that by 2020, more than five million jobs will be lost to robots. The WEF specifically mentioned the fourth industrial revolution in the report title and emphasized that the movement will be unprecedented in "developments in genetics, artificial intelligence, robotics, nanotechnology, 3D printing, and biotechnology."[15]

It isn't only those displaced by machines who must claw their way into shrinking numbers of job positions, but the generations that are rising or will rise in the face of the automation takeover. Those who are not capable of designing, operating, working with, or maintaining the armies of advancing robots could be facing a very uncertain future. Only those who learn to dance with machines will be able to coexist with them.

The main shift in human jobs at the moment lies in such areas as software developers who design new robotic machines, mechanical engineers and maintenance hands who assemble the machines and keep them operating, and data analysts who trace the results and report on the success or failure of deployed machines. Although such job positions tend to pay more and provide bright futures to those qualified to perform them, it is unlikely they will offset the numbers of jobs being lost to robots. An economic paper published in March 2017 revealed findings that as many as six workers lost their jobs for each robot deployed in the US manufacturing industry.[16] And as we shall see, the fields and job positions that are able to be filled by robots are expanding rapidly.

ROBOTS AMONG US

The biggest reason we must accept working with robots is because—well—they're already here. In August 2017, the Brookings Institute® released a study that showed in which US states robots were being used the most. Robots are by no means close to taking over the country, but they are definitely making their mark—particularly in the high-manufacturing states located in the Midwest and South. The bulk of robots being used are defined by Brookings as "automatically controlled, reprogrammable machines" capable of replacing labor in a range of tasks.[17]

The report revealed that large clusters of robots are being utilized in cities recognized mainly for harboring factories that produce automobiles, although other types of plants are using industrial robots as well. Michigan led the pack with almost twenty-eight thousand operational robots, with Detroit having one of the highest robot-human ratios of 8.5 robots for every 1,000 workers. The metro area of Elkhart-Goshen, Indiana, walked away with the ratio award, having 35.6 robots for every 1,000 workers![18] Rounding out the top ten were #3 Grand Rapids-Wyoming, MI (6.3 robots per 1,000 employees); #4 Louisville/

Jefferson County, KY-IN (5.1 per 1,000); #5 Nashville-Davidson-Murfreesboro-Franklin, TN (4.8 per 1,000); #6 Youngstown-Warren-Boardman, OH-PA (4.5 per 1,000); #7 Jackson, MS (4.3 per 1,000); #8 Greenville-Anderson-Mauldin, SC (4.2 per 1,000); #9 Ogden-Clearfield, UT (4.2 per 1,000); and #10 Knox-ville, TN (3.7 per 1,000).[19]

The number of robots being applied to manufacturing operations is impres-sive, but even more interesting is that robots are on the verge of leaving factory floors and entering more mainstream venues. They are also making strides toward greater mobility, including the use of legs, and, combined with higher artificial intelligence levels and adaptability, have the potential to perform more humanlike tasks. Patient care, customer service, and office support are all in the crosshairs of robot designers and investors.

Once robots become sufficiently advanced in autonomous problem-solving and mobility, they will be able to complete such tasks as delivering packages; setting appointments; taking inventory and stocking shelves; providing secu-rity; and assisting with physical therapy, medication administration, and per-sonal hygiene. Most of these activities are already in the testing phase: robots are either being designed to complete them or have already been created and are being tested in real business locations.

ROBOTS IN THE HOME

The many conveniences that robotics technology offers are set to invade our homes, bringing with them a vast host of enjoyable and pleasant gizmos and gadgets to make our lives better. The move toward smart homes is already in full swing, with a variety of products hitting the market as well as continuously being developed. Home automation is rapidly providing occupants with control systems that monitor, calculate, or respond to gathered data, programmed input, or voice commands. Smart devices exist that adjust lighting and environmental temperature as well as provide surveillance and automatically lock doors and windows. While you're kicked back and relaxing, flip on your Roomba® vacuum cleaner and let it do the dirty work while you enjoy a good book on your iPad® or other entertainment from a smart device.

Actual robotlike creations are already in the works. Tēmi® is a combination personal assistant, DJ, and media hub that will follow you around and respond to your requests and voice commands. This interactive robot brings advanced AI experiences into the home, combining television, audio, and internet technology into one intuitive and fun product. Google Home® and Amazon's Echo® line can be placed nearby, awaiting the chance to answer questions, adjust environmental controls, play your favorite music, and more. Smart ovens can cook your meals to perfection and will notify you through your smartphone when they are ready; smart refrigerators come with internal computer systems that allow you to leave messages for family members, track inventory, and place delivery orders for items that have either been specified or are in short supply; and smart electronics can fill your living spaces with pleasurable and educational music, videos, programs, and social network interaction.

ROBOTS IN THE OFFICE

Since a large number of people work in an office atmosphere, it is worth taking a look at how advanced technology is having an impact on office life. There is no doubt that technology simplifies a variety of work-related tasks, and the office environment is not lacking in such opportunities. Technological advances in computers, communication systems, presentation solutions, and more make work projects easier to accomplish and of better quality. These functions, in turn, improve working conditions as well as employee comfort levels and attitudes.

The latter point of maintaining a high morale is an important one, since we spend the majority of our time at the workplace. One recent study revealed that happy employees are 12 percent more productive than the average employee, while unhappy ones are 10 percent less productive than the average employee.[20] Financial incentives were found not to be all that motivating, as simple offerings that relieved work stress tended to prompt productivity over mere financial gain. A seven-hundred-person study conducted by the University of Warwick and the Social Market Foundation® found that snacks and a ten-minute comedy clip were enough to boost productivity by 12 percent on average, with spikes of 20 percent.[21]

Apparently, the effort to make and keep workers happy is well worth it. Megacompanies such as Microsoft®, Apple®, Google®, Facebook®, and LinkedIn® invest a great deal in order to make employees happy. Facebook, for example, boasts the largest open office in the world, offering their talent-laden workers large, open, cubicle-free spaces where interaction is encouraged with the goal of increased creativity and group collaboration. Workers are also given laptops to keep them mobile; libraries, game rooms, and quiet spaces to incubate ideas and relax; flexible schedules to encourage peak performance; and micro-kitchens to grab energizing snacks. Google offers a similar work environment, which has produced an increase in employee satisfaction of 37 percent.[22]

The transition to a new and improved office model is made possible because of advances in technology. Wireless computers, phones, and charging devices allow workers mobility to perform tasks more efficiently from areas that provide greater comfort or are better fits for the work involved. Work flexibility is a key factor in boosting employee satisfaction, creativity, and productivity, and wireless and other forms of technology have expanded such options for companies and their workers. People are now able to work from home using technologies such as work-tracking systems and videoconferencing. Others can use mobile devices to schedule and attend meetings, communicate with clients and staff, send messages or packages, and more while on the go.

Smart sensors are another form of technology that creates a more comfortable and efficient work environment. Through precise control of lighting, cooling, and heating systems, these can save companies as much as 50–70 percent on energy bills. Advances in sensor technology are allowing employees to tailor their workspaces according to optimum performance needs through such features as temperature and lighting control. Heat and pressure sensors can detect when people are inside certain spaces and provide necessary functions for them or turn lights off or adjust temperature controls when those spaces are empty. Smart building technology can even alert staff when equipment malfunctions or when coffee or snack machines are running low, and can track and reorder inventory.

FORCES RESHAPING THE WORKFORCE

It is evident that as the transformational force of technology presses ahead in the areas of automation, artificial intelligence, virtual and augmented reality, quantum computing, and more, the overall structure of the workforce will follow suit and be reshaped as these developments grow in influence. This restructuring of how, where, and even why work will function in the future is due to several powerful shifts caused by megatrends that are in the process of creating major redistributions of wealth, power, opportunity, and competition on a global scale. In the midst of such radical changes, business owners and operators as well as individuals must recalculate their purposes, models, and mandates in order to survive. Adjustments become more critical as these megatrends force society in all of its forms to be reshaped accordingly, and as adjustments are made, the workforce of the future is transformed into something as yet unknown.

There are five megatrends that are responsible for major shifts in the future of the workforce, which should be considered in order to have a clearer vision of where we are headed.

1. ADVANCING TECHNOLOGY

Rapid advances in technology development are not only taking place but are set to move into exponential growth, eventually reaching a tipping point of singularity (something explained further in upcoming chapters). The depth, breadth, and speed at which technological advances are occurring are causing dramatic changes in the workforce today and will increasingly do so into the near future.

Changes are already taking place in the number of jobs available as well as the kinds of jobs being created and offered as demand for more technologically savvy positions increases. In many areas, the number of available jobs is shrinking as robots and other forms of technology take the place of humans. On the other side of the coin, however, the advancement of technology is also creating many jobs that either didn't exist prior to the current industrial revolution or only existed before in small numbers. Many workers who are finding their jobs threatened by technology are crossing over to other careers via extracurricular training, and younger generations coming up under the umbrella of technology are making educational choices that better fit with rising workforce demands.

2. SHIFTS IN DEMOGRAPHICS

The majority of the global population is in the midst of an unusual explosion with a majority of regions experiencing the aging of their people. This is leaving companies in a precarious position of figuring out how to meet production, distribution, service and other relevant demands with a shrinking workforce. People are simply living longer and having fewer children than the generations before them—much of which is due to a surge in technological developments in various areas. Longer lifespans and less "new blood" to fill growing business bases are having a noticeable effect on such areas as talent pools, business structures, educational endeavors, healthcare provision and pension costs.

The issue of aging and other shifts in global demographics are main drivers in automation, AI and other forms of technology development as growing swaths of society race to meet workforce shortages. Many companies and organizations are already turning en masse to reeducation and retooling programs in order to upgrade existing employee resources, since drawing new talent is increasingly more difficult and will only become more so going forward—at least in the short term.

3. VAST URBANIZATION

The trend of large numbers of people moving from rural to urban settings that was set by the first industrial revolution and was followed during the second and third industrial revolutions continues in earnest into the fourth. The United Nations has estimated that approximately 54 percent of the global population was living in cities in 2016. They further projected that large urban areas (those housing half a million or more inhabitants) will contain 60 percent of the world's population by 2030, which equates to roughly two out of every three people.

They go on to report that there will be 662 cities having a population of at least one million by 2030, up from 512 in 2016. Megacities, those housing more than ten million people, will climb from thirty-one in 2016 to forty-one by 2030.[23]

Technology is a main reason for such major shifts to urban dwellings. Not only are greater numbers of jobs produced in cities as both a direct and indirect result of mechanical and computer advancements, but social and living aspects are also being improved through technological advancements, offering city dwellers more comforts as well as opportunities. Another driver in the urbanization shift is due to advancements in technology, which are taking over rural operations that once required large numbers of humans. Agriculture, animal husbandry, conservation, and other areas considered rural are seeing larger numbers of machines developed to fulfill tasks.

4. GLOBAL ECONOMIC POWER SHIFTS

Amid all these changes, another shift is occurring in the area of economic power. High-tech development has rapidly enlarged globalization, which in turn has allowed developing nations to catch up with, and in some cases bypass, developed countries. Those developing countries that possess an available pool of working-age people and are active in promoting educational opportunities have a further advantage over developed nations with a largely aging population. Such advantages tend to draw investors and new businesses, which increases the economic leverage from developed to developing countries.

Nations already enjoying a developed status face further challenges as outdated systems, job loss, middle-class erosion, and wealth disparity as well as population aging take their toll. Developing countries are often more zealous in their efforts to embrace new technology and provide educational and other improvement factors to their citizens. While developed countries put a lion's share of their efforts into restructuring current systems, those nations on the rise tend to go straight to the most current forms of technological development, giving them a competitive edge over larger nations.

5. RESOURCE LIMITATIONS AND ENVIRONMENTAL CHANGES

Yet another megatrend that is responsible for major shifts in the workforce today is the depletion of limited natural resources and an upturn in extreme climate change. These two factors are closely related, as increased use of fossil fuels has (at least) added to extremes in weather patterns we have witnessed over the past several decades. Harsh weather patterns—e.g., droughts, flooding, high winds, heat, and cold—are wreaking havoc on rural areas, leading to mass failures in crops, forestation, and the related ecosystems. Many people simply cannot afford such major or frequent losses, so they move to cities where making a living is more balanced and secure.

Although the shrinking of natural resources and the escalation of extreme weather have their dire consequences, they also offer benefits to those seeking new opportunities. Experts are turning to advancements in technology to deal with both problems, which is creating a growing number of better-paying specialty jobs in the fields of alternative energy development, product design, engineering, waste management, food production, and water purification, to name a few. As an example, there is a desperate need for clean water to meet population requirements and greater sources of energy to power the machines and devices that are filling our factories, businesses, cities, streets, and homes. It is estimated that by 2030, there will be a 50 percent increase in energy needs and a 40 percent increase in clean water requirements.[24] Massive increases in resource demands will contribute to a rapid and major restructuring of the workforce as we move forward.

It is clear that various forces are in play that can and will transform the workplace as well as societal structure as a whole. As we race headlong into the future, such megatrends and other transformational events and trends should be both acknowledged and acted on in order to maintain our personal competitive edge during such turbulent changes. Shifts in the overall structure of the workforce are set to occur, and there is nothing we can do to stop them. Since outside forces cannot be halted, it is best to take a proactive stance and make the necessary changes in those areas we can still control, such as education, alternative career building, networking, and living location.

CHAPTER 4

ARTIFICIAL INTELLIGENCE: FROM SCIENCE FICTION TO REALITY

People have always been intrigued and fascinated by what the mind can imagine. It is the cornerstone of the multibillion-dollar entertainment industry that churns out a continuous stream of television shows and movies that captivate our attention and feed our longing for the seemingly impossible. It's a great way to escape our day-to-day troubles, worries, and fears.

Indeed, most of us grew up watching science fiction movies and television shows that left us in awe of the seemingly magical gadgets they employed to navigate star systems, fight monsters or aliens, or provide life necessities and creature comforts on a mere whim. We marveled at the phasers, communicators, transporters, and instant food delivery systems of *Star Trek*; lightsabers, R2-D2, C-3PO, and BB-8 in the Star Wars saga; the powerful abilities of the Bionic Man; the artificial intelligence and keen trickery of HAL in *2001: A Space Odyssey*; and the militant crime-fighting capabilities of the cyborg law enforcer of *RoboCop*. Much of the appeal of such shows was due to the technological devices that were thought to be fantastical and well out of the reach of reality. They inspired our minds with wonder and possibilities.

The true wonder of it all, however, is that a large number of those fantastical items we have enjoyed while watching movies and television shows have become or are quickly becoming realities today. Items created in the minds of writers, directors, and producers have escaped the pages and screens to which they were confined and have infiltrated and affected our lives and our world. We now have or will soon have machines that can be programmed to accomplish some of our wildest expectations, and many of them are driven by their own (artificial) intelligence. The imagined elements of science fiction have indeed become the elements of our reality.

MANIFESTATIONS OF THE IMAGINABLE

It is interesting to glance back at old movies and television shows and compare the technology that was near-total fantasy at the time. It's as if writers and directors were gazing into a crystal ball and recreating their visions to entertain the world. Every generation has grown up with such fantastical stories and looked to turn fiction into reality. We saw an example of this as space travel transformed from a concept of science fiction into something that is now commonplace. In 1865, Jules Verne published his remarkably accurate account of astronauts traveling to the moon—a feat that would occur just over a century later on July 20, 1969. Since then, countless stories of space travel, life on other planets, intelligent robots, and much more have filled our minds with wonder and ideas.

It is well worth accompanying the ghost of entertainment past and taking a glimpse at the gizmos and gadgets it shows us in order to examine some examples of those devices and abilities portrayed in science fiction movies. Seeing how they have wonderfully manifested into our present time is just as entertaining as when they were revealed to us in mere stories decades ago. Certainly, some investors were inspired to make these visions reality.

Following are a few of those sci-fi wonders that have miraculously escaped from the page and the screen to become real objects today. Decide for yourself if some strange magic was at play.

MOBILE PHONES

The original *Star Trek* television series that ran from 1966 through 1969 was filled with fictional apparatuses that have eerily become real operational items in our world. It is as if Gene Roddenberry operated as some high-sage wizard who effortlessly foresaw futuristic technology and applied it to his intriguing tale of space exploration.

At the time this tale of deep space exploration occurred, mobile communicators that would operate from various ship locations or even from the surface of explored planets were already happening, even with the early space program. Therefore, one of the main objects used in nearly every episode of *Star Trek* was the communicator. The handheld gadget was similar in size, operation, and appearance to today's smartphones. Captain Kirk could simply flip it open and glean lifesaving, distant advice from Spock, Scotty, or Bones. Although communicators were impressive at the time, smartphone technology is everywhere today. Scientists are even working to emulate the Enterprise's transporter by "beaming" objects from one place to another.[1]

VIDEOCONFERENCING

Another concept that has become part of our lives is videoconferencing. The concept of talking with another person on the phone while being able to see her face through a video screen has been around for a while. Movies like *2001: A Space Odyssey*, which hit theaters in 1968. promoted the concept, as did numerous other sci-fi stories. This had the effect of inspiring hope in such technology, since both pre- and post-WWII experimentation existed—although with less than hopeful results. For example, Bell Labs rolled out a working videophone system in the 1970s that was overwhelmingly rejected.

It wasn't until the computer revolution emerged, that videoconferencing began to emerge as a viable option. The earliest versions were QuickCam®, released in 1994, and CU-SeeMe® software, released in 1995, but neither took off, remaining mostly used by businesses. Widespread use of videoconferencing didn't gain massive popularity until Skype® was released in 2003, offering free crossplatform service to internet users. Such technologies for videoconferencing continue to evolve, with collaboration tools like Microsoft SurfaceHub® that enable videoconferencing and the use of a shared display area for remote workers to write and share information. Amazon's Echo Show® brings voice-activated video calling to your home.

HOVERBOARDS

Ever since Marty McFly hopped on a futuristic hoverboard to make an exhilarating escape in the 1989 release of *Back to the Future II*, kids (both big and small) have drooled over the possibility of joining him. That technology is finally surfacing, and it arrived, appropriately enough, in 2015—the year of the "future" depicted in the movie. Although the technology is still limited, big strides are being made to bring Marty's hoverboard into reality. Both the Hendo Hover® (by Arx Pax®) and the Lexus® board (by the carmaker of that name) have utilized electromagnetic technology to create real-life hoverboards that create and float on a magnetic field that is from one to several inches high. The big drawback? They both require special surfaces to operate correctly. Surfing through the concrete streets and over water fountains like Marty McFly will have to wait, or be experienced in a premade setting.

There is another solution from which to glean craved adrenaline rushes. The Omni® hoverboard incorporates drone technology to propel riders through the sky. Like Spider-Man's nemesis, the Green Goblin of Marvel Comics® fame, the Omni can lift its rider's feet off the ground and move over various surfaces. Its Canadian inventor, Alexandru Duru, demonstrated the effectiveness of his creation by breaking a Guinness World Record—flying the Omni over Quebec's Lake Ouareau for a distance of 905 feet, 2 inches, and shattering the previous record of a dismal 164 feet for such a flight. The Omni showed its viability as a viable hoverboard product and is now due for commercial release.

It isn't only hoverboards that are being researched and developed. Remember the speeder bikes (jumpspeeders) used in the Star Wars movies? Hoverbikes are also on the drawing board and may be zipping around roadways soon. The Hoverbike P1® (by Malloy Aeronautics Ltd®) already exists for beta testing. The P1 is the size of a small car and can either fly autonomously or carry a rider. It has a 130-kilogram payload capacity (286.6 pounds), can reach sixty miles per hour, and can climb to a height of ten thousand feet.

DRIVERLESS CARS

Automated driverless cars have inspired audience minds in television shows and movies for decades—from the autoresponding Batmobile on the television series *Batman* (1966–1968) and the AI-driven KITT in *Knight Rider* (1982–1986) to the pod cars in the movie *Logan's Run* (1976) and the automated taxi with robot Johnny Cab that chauffeurs Arnold Schwarzenegger around in *Total Recall* (1990). Today, smart cars have captivated our attention as a sign that the "future" has indeed arrived.

Driverless cars are a rapid up-and-coming trend that has many companies racing to attain the edge that puts their versions on roads en masse. Although there are many plans and prototypes in the works, some of the most ambitious are becoming realities. You will find examples of real uses for driverless cars in a later section of this chapter, "AI and Robotics Today: Companies Paving the Way."

INVISIBILITY

The ability to become invisible at will is a desire that extends back to ancient history. Greek mythology, for example, told of a "cap of invisibility" owned by Hades that made the wearer disappear in stealthlike fashion. Norse mythology told of a similar helmet, known as Tarnhelm, and the Welsh recounted a tale in which Caradog ap Bran was murdered by the invisibility-cloak-clad Caswallawn.

Invisibility has been portrayed in many books and films. The main character in the English fairy tale "Jack the Giant Killer" was gifted with a coat of invisibility by a spared giant. German fairy tales "The Twelve Dancing Princesses" and "The King of the Golden Mountain" mention similar items. The Harry Potter book series, authored by J. K. Rowling, had the wizardry student Harry utilizing a cloak of invisibility to move about undetected. The main character in the Lord of the Rings trilogy (both the books written by J. R. R. Tolkien in the 1930s and 1940s and the films released in 2001, 2002, and 2003) relied on both a cloak of invisibility and a ring to avoid detection by evil foes.

In addition to all the above magical paths to invisibility, the entertainment world doesn't lack for technological cloaking. In both the television shows and movies of the Star Trek universe, cloaking technology is used to hide ships and other large objects. According to the series, cloaking technology was originally developed by the Romulan alien species. It seems aliens are expected to utilize such technology. The 1987 movie *Predator* (as well as its 1990 sequel) demonstrates both the effectiveness of using such cloaking technology and the challenge, from the human prey's perspective, of overcoming it.

The good news is that real cloaking technology is very much alive, well, and advancing among humans. Such devices operate by distorting light, radar, and other electromagnetic waves in order to make the detection of covered objects more difficult for the naked eye as well as certain devices. Although the technology is still not completely effective for commercial use, scientists are racing to develop working models to render humans and objects, even those as large as airplanes, invisible. Strides in cloaking technology are being made in the military arena, but it is difficult to know how far advancements go, since such data are classified.[2]

In a more public domain, devices displaying cloaking technology are available that give us a good idea of how such ideas are advancing. The University of Rochester has developed a cloaking system that uses a series of lenses that bend light rays, hiding any object in its path from view. An audio cloaking device has been developed by scientists at Duke University that uses sound waves to hide objects. Cloaking technology is primarily made from metamaterials—human-made elements that bend and deflect light rays so that they do not hit the target, thus making it undetectable to the human eye. However, Imperial College London found a way to hide objects using these metamaterials.[3] They took it a step further by using space-time jumps to hide objects and events, in the same way that video feed is deleted from surveillance footage.[4] Another method, projection cloaking, hides objects that are clothed in a special reflective material that appears invisible under the beams of a projector.[5]

FLYING TAXIS AND CARS

Fans of *The Fifth Element* may remember Bruce Willis as the flying taxicab driver Korben Dallas. Of course, this is not the only movie with flying cars and taxis, but it is one that stands out. Flying taxis are actually a reality. Dubai has promised to be the first city in the world to have autonomous flying taxis available for its citizens and guests. In fact, Crown Prince Sheikh Hamdan bin Mohammed has taken an early test ride in one.[6]

AI AND ROBOTICS TODAY: COMPANIES PAVING THE WAY

There are some very important and attractive reasons why AI and robotics are being so aggressively pursued and employed in companies spanning a wide swath of industries. Robots significantly and positively affect such areas as safety, production speed and efficiency, ROI and the overall bottom line, and other critical factors that lead to healthy competitiveness and growth. Because advanced technology offers such promising benefits for both the companies that develop products and the people who use those items, many in the business world are rushing to be the first to offer robotic devices and systems that will meet those positive goals.

Many of the corporate giants of today, as well as new start-ups following their example, are turning to open office spaces that eliminate cell-like enclosed squares and cubicles in order to encourage and maximize creativity, productivity, idea sharing, and other community-enhanced values. However, such an environment has a major drawback—the loss of privacy, space, and quiet, which are necessary aspects in the creative and productive formula. Therefore, a merger of the two ideas is being experimented with that utilizes advanced technology in the form of virtual workstations. The VR displays being used are contained in a headpiece that is complete with all the data platforms required to perform work tasks, such as search engines, word processing and spreadsheet software, and email and communication programs. The only physical element binding the worker to the space is a camera digitally tied to a keyboard. Everything else dwells in the realm of virtual existence. Because such VR systems operate almost entirely in a virtual world, they can be accessed from home or other places besides the office, which significantly cuts company overhead by reducing the need for energy, communication hubs, and other expensive requirements.

Even though there are still a myriad of unanswered questions and shadowy speculations concerning the influx of robots into society, some companies are deploying new advancements and are acting as guinea pigs in the global experiment. Many technologies are in their infancy, while others are still only elaborate and hopeful ideas on the drawing board. Some new technologies fail, some old versions become outdated, and still others find their place in an increasingly technological world. Will an aggressive drive to embrace and implement robots and other new technologies in the workplace produce the golden egg that gives a company that desired advantage over its competitors?

The following companies aren't waiting by the sidelines to find out if robots are the future—they are taking the lead in discovering the results for themselves. This is by no means a complete list, but it clearly shows that a major transformation of the workplace is well under way.

AMAZON®

Amazon specializes in electronic commerce and has gone from a humble online bookstore to a massive global, internet-based retailer since its start in 1995. The development and implementation of high-tech systems have been responsible for the lion's share of Amazon's success. The importance of pressing forward in this area was made obvious in 2012 when the company acquired Kiva Robotics®. Since that acquisition, Amazon has been busy replacing its human workforce with robots. Presently, the company has more than thirty thousand robots working in its warehouses to anticipate, receive, pick, and deliver orders, and it expects to staff all repetitive jobs with machines very soon. Amazon is a prime example of the shuffling of jobs that is ongoing due to the rising implementation of technology. The online giant employs less than a third of the workforce Walmart® did in 1985, yet the company creates many other types of jobs, such as agency workers and package carriers, on which it is heavily dependent. Over the past sixteen years, for example, more than one hundred thousand UPS® positions have been created due to the uptick in online shopping.[7] Of course, this type of disruption is bittersweet. New jobs are welcomed by those seeking them, but those losing their jobs to automated and computerized systems might not be so joyous.

Robotic technology hasn't only been used to replace in-house workers. Amazon is also developing cutting-edge technology and devices to benefit its customers. A few of its inventions include:

> AmazonFresh®—This service allows customers to order groceries, which are delivered the same or next day. At its beginning in 2007, the service was only available to certain areas in its home base of Seattle, Wash-

ington. However, it began to expand to other cities in 2013 and is now available in San Diego, San Francisco, Los Angeles, New York City, and Philadelphia. And with Amazon recently purchasing Whole Foods®, more is sure to come.

Amazon Fire TV®—This product allows users to either watch live-streamed television or choose from a large selection of shows and movies. It can also be used for gaming.

Amazon Dash Button®—This control device is placed within various rooms of a house where products need to be replenished on a regular basis. When it's time to restock, a button corresponding to a preset item and quantity is depressed, and the selection is automatically ordered and shipped.

Amazon Prime Air®—The company was one of the first to jump on new drone technology once it was approved for use within the United States. Using drones could revolutionize the package delivery industry by allowing faster deliveries and significantly reducing carbon fuel use by eliminating huge traditional fleet numbers. Before delivery drones can be utilized effectively and on a grand scale, Amazon must first over-come some tough hurdles such as government regulations, the need for "drone airports" or docking stations so drones can cover long distances, and automated obstruction avoidance. Regulators in the United Kingdom were found to be more willing to accommodate drones within their airspace than those in the United States and a Prime Air drone made its first official delivery in December 2016 to a Cambridge customer. Services have been expanding in that area since.

CHIPOTLE®

Amazon may be tackling the use of delivery drones on a grand scale, but other companies are utilizing them on a more localized basis. Chipotle and Google® have teamed up to test a drone delivery system called Project Wing®. The testing ground covers the Virginia Tech campus, where burrito orders will be delivered to hungry students from a specially set up kiosk.

Google is especially interested in drone delivery systems that can be used to expedite deliveries and provide supplies during disaster relief efforts, all while greatly reducing fossil fuel usage.

JUST EAT®

Another restaurant is pursuing robot delivery options at ground level. Just Eat is the United Kingdom's largest online ordering and delivery restaurant chain, with more than thirty thousand national locations and sixty-four thousand global restaurants. In February 2017, the company started using self-driving delivery robots in the London suburb of North Greenwich. The driverless delivery devices navigate the roadways using cameras and GPS technology.

PIZZA HUT®

Everyone loves pizza, right? Well, it appears that the Chinese want to mix their Pizza Hut delicacies with robotics. The outlets throughout China utilize humanlike Pepper® robots created by SoftBank® that can answer questions, take orders, and process MasterCard® payments.

It would stand to reason that robots that can perform those kinds of tasks could also make the deliveries to the tables of waiting customers. Apparently, the Chinese think so, because a Pizza Hut in Shanghai is experimenting with such an idea. The store uses friendly robot waiters to greet customers, escort them to an available table, and take and deliver orders.

DHL®

Like Amazon, the logistics giant DHL is turning to robots to more efficiently and effectively process customer orders. The company purchased several of the smart robots Baxter® and Sawyer® in 2017 to assist warehouse workers with such copacking and preretail tasks as assembly, packaging, and kitting. DHL plans on making steep investments in robotic technology throughout 2017 and 2018 to support staff and enhance productivity.

CAMBRIDGE INDUSTRIES GROUP®

As mentioned, China is paving the way in both the development and deployment of robotics, and CIG is a prime example. The Shanghai-based company is a major global supplier of telecom equipment that is aggressively applying automated systems to its operations.

UBER®

The up-and-coming ride-sharing company Uber is in the process of testing its version of self-driving cars. Although it has run into regulatory resistance in its chosen test city of San Francisco, Uber is pressing toward its ultimate goal of only using robot drivers.

Driverless technology is on the fast track of development, and our roads and highways are sure to be filled with robotic drivers in the very near future. There are simply too many companies jockeying for position in the market that are sure to overwhelm and eventually convince regulators to allow them. For example, Tesla® has announced that its line of driverless cars will be ready to deploy by September of 2018. Other companies in hot pursuit of driverless technology are Ford®, General Motors®, Toyota®, Honda®, Hyundai®, Volvo®, Nissan®, Audi®, BMW®, Daimler® (producer of Mercedes-Benz®), Baidu® (car manufacturer based in Beijing), LeEco® (Chinese tech company), Bosch® (designer of driverless technology), PSA Group® (European car manufacturer), Google, Faraday Future® (specializing in electric car designs), and Apple® (which has opted to focus on autonomous software design over pursuing self-driving cars).

TESLA

Although Tesla (mentioned above) is pursuing driverless technology, the company headed by tech giant Elon Musk is also utilizing robotics to construct its creations. Currently, Tesla's huge Gigafactory® in the Nevada desert is run mainly by machines with only a small crew supervising operations. The goal is to have the Gigafactory as well as other plants operating on 100 percent robotics in the future.

CAPITA®

Capita is a major player in the business world, sitting on the Financial Times Stock Exchange 100. The UK-based company is currently investing in both robot and cobot (robots working alongside humans) technology in what it has announced as a cost-cutting measure. The company's immediate goal is to replace two thousand of its human employees with robots.

WALMART

This department store megagiant has consistently utilized cutting-edge technology to streamline its operations and workforce, and it doesn't seem to be slowing down. Walmart is in the process of expanding its use of automated technology beyond its distribution centers. For example, the company is rapidly moving to replace traditionally human-operated jobs in its accounting and invoicing departments with robots.

But that's not all. Walmart has patented a system of self-operated shopping carts, which it intends to place in its thousands of stores. Instead of requiring staff to retrieve shopping carts left in aisles and parking lots, the computerized carts would run and return on their own. These smart carts are also designed to move certain types of items; take inventory; clean up trash; interact with customers; and scan, retrieve, and deliver products.

OCADO®

More than a thousand robots have already been deployed in the warehouses of Ocado to supply products to its online customers. However, the British company is pressing ahead with automation technology, employing assistance from several major universities for the development of robotic humanoid technicians that will be able to handle complicated maintenance tasks.

LOWE'S®

Lowe's seeks to keep its edge in the hardware supply store industry by turning to customer service robots. The OSHbots®, or "Lowebots®" as they've been tenderly dubbed, are able to assist customers via verbal communication, touchscreen menus, or 3-D scanners that can identify items customers have and need to purchase. Lowebots are self-driving and can lead customers to desired products, take store inventory, and converse and answer questions in various languages using multilingual software.

ADIDAS®

German sportswear company Adidas has launched a fully automated manufacturing facility dubbed Speedfactory®. The factory, located in Ansbach, Bavaria, produced a streamlined running shoe it unveiled in 2016. The sneakers are created by robots with very little human assistance, using ARAMIS® technology to create a better-performing shoe based on musculoskeletal data. Adidas's second Speedfactory is located in Atlanta, Georgia.

CARRIER®

Many of you may recognize Carrier as the company that then president-elect Donald Trump convinced to remain in the United States instead of moving to

Mexico. The news of saving eight hundred jobs made headlines. However, the catch-22 is that although the air-conditioning manufacturer agreed to the deal, it is using the Indiana site as an investment springboard where it can develop and deploy robotic systems that will eventually replace or eliminate many of those eight hundred positions.

NESTLÉ®

Nestlé is gambling on Japan's love of technology and science to increase sales of machines that disperse its popular Nescafé® product. The company has "employed" a humanoid robot to act as its sales representative, which it is utilizing in more than a thousand home appliance centers and stores throughout the country. The robot, known as Pepper and developed by Softbank, uses recognition technology to identify and respond to human facial expressions and emotions. Besides interacting with customers, Pepper robots are also programmed to demonstrate products and assist with making purchases.

ING®

Banks are included in the barrage of industries seeking to use robotic technology in order to cut employee expense and improve performance. The major Dutch chain ING has already turned to robots for assistance with trimming costs. ING is in the process of eliminating 5,800 employees, the majority of whom are full-time, with an additional 1,200 being relocated or moved to other positions. ING seeks to centralize its human resources, finance, information technology, and risk-management operations into one platform.

The company has its eyes set on investing around €800 million in digital banking technology over the next half-dozen years. Although experiencing resistance to the change, particularly by unions, the company stresses that it must modernize in order to compete with tech-savvy giants that are rapidly implementing robotic technology.

SHISEIDO®

The Japan-based company Shiseido has begun using industrial humanoid robots in its Kakegawa factory to assemble cosmetic products, which is a first in the industry. Two cobots work alongside one human employee, who is basically responsible for inspecting the products and ensuring quality control. The company seeks to more swiftly adjust to changes in the market by using robots and humans together to increase flexibility.

FOXCONN®

The Chinese-owned company Foxconn, which manufactures devices for Microsoft®, Samsung®, and Apple, has an aggressive plan to create 100 percent robot-operated factories in the near future. It has already started the process by replacing some sixty thousand workers and has designed a three-stage plan to achieve this goal. Phase one deploys robots to positions that are either dangerous or undesirable for humans. Phase two involves streamlining operations to further reduce the need for both humans and robots. Phase three seeks to have almost fully automated factories, requiring only a skeleton human crew to perform such tasks as testing, inspection, logistics, and certain production jobs.

MARRIOTT® HOTELS

Marriott is the hotel chain leading the charge in the world of robotics. The company's Belgium Ghent Marriott deployed a humanoid robot named Mario® in June 2015 to assist with such tasks as checking in guests, giving speeches and PowerPoint® presentations at meetings and conventions, offering restaurant menu advice, and ensuring rooms are cleared after checkout. The biggest asset of having Mario around doesn't seem to be his handiness at handling hotel tasks. The robot can communicate in nineteen different languages and has become a big hit with visiting guests as well as locals, especially children. Local schools often schedule field trips to visit Mario.

NISSAN

Although auto manufacturing companies have been utilizing robotics for years, Nissan has pushed the envelope. Its factory located in Sunderland, England, is 95 percent automated and extremely efficient. The plant's robots annually produce five hundred thousand vehicles and can build one of its Qashqai® SUVs in a mere 8.5 hours.

EVERWIN PRECISION TECHNOLOGY®

The Chinese tech giant Everwin Precision Technology, located in Dongguan City, is well on the way to creating a factory run almost exclusively by robots. The company uses automatons for a majority of its machining and transport operations, reducing its employee numbers from an original 650 to a current 60. The goal is to drop that number further, to a mere twenty employees tasked with overseeing the army of robots.

Is the switch to robots over humans worth it? The company's results definitely supply a resounding yes. Robots produce nearly three times more than their human predecessors, pumping out twenty-one thousand items compared to eight thousand, which is a 162.5 percent increase. Product quality has also improved from a human-defect rate of 25 percent to a robot-defect rate of 5 percent.

FIDELITY INVESTMENTS®

Traditional financial advisors are also coming under the gun of advanced robotic technology. Millennials (often defined as those born after 1980 up to the twenty-first century) are particularly the ones seeking more advanced methods of financial management and are turning in droves to what have been dubbed "robo advisors."

In order to keep up with an increasingly tech-advancing competition, Fidelity has launched its own robo advisor system called Fidelity Go®. Although the digital operating model is designed to meet the needs of first-time investors, Fidelity intends to further modify the technology to handle more advanced core investment capabilities that millennials and those coming after are seeking.

ZARA®

Zara is a Spanish apparel company that, like so many other businesses, saw its future doomed to failure unless something was done to give it an edge in the industry. That something was automation. While many of its competitors were turning to subcontracting to factories in cheaper locations, Zara chose instead to invest in its home companies. It constructed fourteen new factories that were predominantly operated by robots that could work around the clock. These robots cut and dye material, producing what is known as "gray goods"—fabric that is ready to be made into finished products.

The robot-produced gray goods are shipped to a network of several hundred shops that complete the finished works. This rapid approach to mass-style design allows Zara to quickly identify top fashion sellers who can then distribute to stores within a few days. As the saying goes, "The proof is in the pudding"; Zara's parent company, Inditex®, gleaned an additional 10 percent in revenue growth at the end of fiscal year 2012, which equated to $19.15 billion. The results of its automated "just in time" system, which is patterned after the one used by Toyota®, has caught the attention of other clothing manufacturers, which are implementing the same or similar systems.

THE RISE OF ROBOTS

We have looked at how imagined devices created for our entertainment have become realities in our world today. We have also considered how various companies are operating on the cutting edge of robotics technology to reduce costs, boost production, improve service, and provide products that appeal to their customers. These are all important and impressive examples of how robotics and AI are being developed, honed, and deployed in order to meet those goals.

However, the operational boundaries of intelligent bots are expanding, and they are increasingly found performing jobs that were believed to have been reserved for humans. As technology advances, more robots will be able to perform increasingly complex tasks and, in turn, replace more human workers who rely on those jobs for income. The rise of robots no longer dwells in the fantastic world of science fiction but is happening right here, right now.

ROBOT LAWYERS AND PARALEGALS

Most people would think that providing sensitive legal advice and services is confined to real breathing, thinking humans who have spent years at a university learning the craft. However, such a notion is being challenged as robot lawyers are being introduced. As with the majority of AI assignments, the bulk of tasks are routine and mechanical, such as reviewing and retrieving data, but the hope of robots filling more complex positions is on the foreseeable horizon.

In our first example, a Facebook® chat bot dubbed "DoNotPay®" was designed by Joshua Browder, a twenty-year-old British student of Stanford University, to be a legal representative. The robot lawyer initially helped some 250,000 people challenge parking tickets, with more than 160,000 successfully overturned. Browder has since tweaked the program to perform other legal tasks such as obtaining delayed-flight compensation; providing legal advice for HIV patients; and assisting refugees with claiming asylum in the United Kingdom, United States, and Canada. The DoNotPay bot operates through the interface of Facebook's Messenger program, where it converses, asks questions, and records responses of interested people. Lawyers gladly welcome the potential offered by the legal bot assistant. In the case of immigration lawyers, for example, the lion's share of their time is used on trivial activities such as filling out legal forms. Bots, such as the one designed by Browder, have the potential to release lawyers from such tasks so they can focus more intently on case complexities.[8]

It isn't only social media sites implementing legal robots. Lawyers from every size and type of organization are also adding AI bots to their workforce, which is set to dramatically transform the industry. LawGeex® is one AI platform being used that compares contracts in a database and creates one specifically tailored to the present need. Another AI platform, eDiscovery®, conducts searches of filed documents in order to retrieve data required to pursue various forms of litigation. Are such bots more efficient than their human counterparts? Apparently so, as the large international law firm Reed Smith discovered. They tested an AI system by RAVN Systems® known as RAVN ACE® by having the robot review hundreds of pages of documents. At first, there were some mistakes, but lawyers added certain necessary data and results improved. The end result? The AI bot was much faster, taking mere minutes to process tasks that normally took humans days to complete.[9]

A growing number of megafirms are jumping onboard the AI bot wagon. Latham & Watkins®, Denton®, and other giants have turned to an AI system called Watson® by IBM® and ROSS Intelligence®. Although Watson is still in its infancy, the system is already saving law firms mountains of research time— 20–30 hours per client. It is estimated that 70 percent of legal costs are incurred through search-and-review tasks such as combing through documents and reading emails. If robots take over such tedious assignments, significant savings could be realized.[10] The technology has already made an impressive jump from a system built on the Watson platform to one simply called "Ross®." This intelligent attorney software has been deployed by law firm Baker & Hostetler® to take care of the bankruptcy branch of their practice. The AI platform can perform more sophisticated tasks such as communicate, answer questions, conduct research, and provide hypotheses and responses. Based on required needs, Ross scans relevant laws, legislation, and case histories to provide fast advice. The AI system also monitors newly applied decisions 24-7 that may affect specific cases and provides notifications.[11]

Like all AI programs, legal bots learn from every experience, which leads to an expansion of their abilities and performance. Lawyers and legal firms stand to gain the same benefits of AI bots as other industries: greater efficiency, lower costs, and faster process times. Those seeking legal advice and representation are also set to benefit. One of the main obstacles preventing ordinary people from receiving good legal assistance is the high cost of securing a lawyer. As robots take over time-consuming tasks, they free up human lawyers to focus on more complex issues, thereby reducing costs and increasing available time to take on additional clients. Growing numbers of individuals and small businesses that were shut out of good legal representation will find they have greater access. Legal AI bot assistants are well on the way to making that happen.

MEDICAL ROBOTS

The roles of medical professionals are also on the verge of experiencing radical transformation as robots rise to take on functions once thought to only be carried out by humans. Although many medical robots are still confined to research and testing, others are successfully moving into roles of action, particularly where masses of information need to be processed or when intricate and delicate maneuvers are required.

The introduction of robots used to complete surgical tasks was first made in 1985 when PUMA 560® successfully inserted a needle into a brain during a biopsy procedure. Then, in 1987, a robot was used to perform a cholecystectomy. A year later, PROBOT® was employed to assist doctors at Imperial College London with prostate surgery. Today, we have a growing list of medical robots showing up to perform tasks in such areas as pediatrics, orthopedics, radiosurgery, and neurosurgery.

IV robots and "robotic syringes" are being used to mix, measure and administer medicine for patients. Many facilities in both the United States and Europe use a robot to measure and mix sensitive medications. For example, chemotherapy requires a precise mixture that if done incorrectly can endanger patients as well as medical personnel. RIVA®, IntelliFill®, and IV Station® are other robots developed for use in medical and pharmaceutical arenas to automatically prepare IV solutions. Precision calculations by robotic syringes and IV robots significantly reduce the possibility of overdose and death, contributing to saving hundreds of lives each year that would normally be lost to human error. One study found that "intravenous (i.v.) medication errors are twice as likely to cause harm to patients as medications delivered by other routes of administration (such as tablets or liquids), according to research commissioned by the American Society of Health-System Pharmacists (ASHP)."[12]

Israeli researchers have designed a microrobot that is only a millimeter in diameter. The ViRob® is used to enter and maneuver through small cavities and vessels in order to locate tumors, place microcatheters, and deliver pharmaceuticals and is steered by an external system utilizing electromagnetic technology. Nanotechnology creates possibilities on an even smaller scale, with minuscule nanobots being tested and used to locate and destroy cancer cells, deliver medications, provide medical imaging and sensing, monitor blood chemistry, and replicate molecules.

One company, Intuitive Surgical®, seeks to utilize intelligent robotic technologies in order to improve surgical experiences for both doctors and patients as well as their families. Their robotic surgical system, known as da Vinci®, allows surgeons to perform minimally invasive operations utilizing the robot's magnified 3-D high-definition camera and minuscule instruments that perform maneuvers more precisely than a human hand. Furthermore, the da Vinci system can be remotely operated over the internet, allowing doctors to perform surgeries from any location around the globe.

In many cases, attending to a patient doesn't stop after treatment. Many people need postoperative and other types of in-house care that require an army of attending doctors, nurses, and aides. Robots are being used to fill those tasks as well. For example, Mr. Gower® is a medical bot that maneuvers through the maze of hospital hallways, elevators, and rooms to deliver needed medications to nurse stations. Doctors are also using remotely controlled robots in some US hospitals to visit patients in intensive care units and take their vital signs without having to leave their stations. Similar robots can execute such tasks as giving diagnoses and implementing stroke management.

A robot dubbed TUG® is a hospital housekeeper and food deliverer that communicates with data systems via wireless internet and maneuvers around hospitals by using an internal integrated mapping system. Sensors allow the TUGs to circumnavigate obstacles, and radio waves are used to enter elevators. These helper robots can carry up to one thousand pounds, allowing them to deliver bed linens and meals. They also have special compartments for transporting medications; these require doctors or nurses to apply their preregistered biometric fingerprint and enter a PIN to gain access. TUGs are self-charging, which allows them to chug away for as long as ten hours and cover approximately twelve miles daily.

As in the legal field, the medical community is receiving robots with open arms. The gap between physician supply and demand is widening at an alarming rate. One report released in 2013 projected a shortfall in primary care physicians of as much as thirty-one thousand by 2025—in the United States alone. Non-primary care physicians in the United States fared worse, with an expected shortfall of up to sixty-three thousand—half of those surgical specialists, and the other half split between medical and other types of specialists.[13] It is anticipated that robots will rise to help meet some of the growing demand for healthcare services.

ROBOT TEACHERS

Robots are blazing a trail into educational venues, and although they haven't yet replaced their human counterparts, they are well on the way to doing so. The replacement of teachers is possibly a driving factor in why the majority of educators are reluctant to bring robots into their classrooms. Most educators insist that robotic roles must be restricted and that teachers must have full control over them.

So far, robots are mostly used as learning tools rather than real teachers, but that trend is quickly changing. Students are much more enthusiastic about the idea of robot teachers, especially those with a humanoid form. On the downside, robots lack (at the moment) emotional responses that help students relate. On the upside, robots do not grow tired or frustrated when pupils make mistakes.

A number of countries are experimenting with humanoid robots in their classrooms, particularly Asian countries such as South Korea and Japan. Robots such as Pepper, I®, Tiro®, Nao®, and IROBI® are being used to teach students English. Telepresence technology is also being implemented to teach English, allowing a teacher to interact with students from a remote location. One robot using telepresence technology has a humanoid design with a display monitor on its chest through which teachers and students can interact. As another approach, Hanson Robotics® offers Professor Einstein®, a small version of the real Albert Einstein. The professor walks, talks, and teaches kids a variety of science topics.

Robot teachers are not only being deployed to the classrooms of children but are also being used to train professionals. For example, SimMan® 3G is a robot that simulates patients in need of medical attention. Students training to become medical professionals are able to experience lifelike emergencies and scenarios in a classroom setting. SimMan provides trainees with realistic symptoms, vocalized distress, and reactions to treatments and medications.

HOSPITALITY ROBOTS

Another industry set to make large strides in the use of robots is the hospitality industry. Panelists attending the Americas Lodging Investment Summit held in January 2017 predicted that room service and housekeeping robots would be the norm in five years. Advancements in technology, decreasing costs and customer expectation were all seen as contributing factors for the projection.

Hotels already experimenting with robots in their facilities give them a big thumbs-up. Robots like Relay® and Botlr® deliver requested food and amenities to guests. When a call comes in to the front desk, the requested item is placed in the bot, the room number is punched in, and the robot makes its way to the destination. The compartment locks, preventing other guests from pilfering items, and then pops open to deliver the goods when sensors detect someone answering the door. Although robots cannot yet climb stairs, they can use elevators and navigate hallways and recognize rooms—all via special mapping software that allows the building's layout to be preprogrammed into the bot.

Meanwhile, the Maidbot®, lovingly dubbed "Rosie" (after the *Jetsons* cartoon's futuristic cartoon maid robot) is busy cleaning and vacuuming hotel hallways, bedrooms, bathrooms, and conference room floors. The makers of Maidbot estimate that each unit can save hotels around $9,000 in various maid-associated costs. Besides relieving staff of meticulous duties and saving hotels a nice chunk of change, hospitality robots like Rosie, Relay, and Botlr are a huge hit with guests, who consider them cute and interesting.

However, most hotels in the western world are not ready to go full robot like one Japanese hotel did in 2015. The Henn-na Hotel®, located near the Huis ten Bosch® theme park in Nagasaki, uses some of the attractions as robot inspirations, such as dinosaur receptionists. Guinness World Records acknowledged the Henn-na (Japanese word meaning "strange") as the world's "first robot-staffed hotel." Receptionist bots speak four languages (Japanese, Chinese, English, and Korean), and porter bots carry luggage and guide guests to their rooms, where facial recognition software gives them access. A small concierge robot occupies each room and turns devices on or off on command and gives information on obtaining food and taxis. Robots also tend to such duties as floor and window cleaning.

The parent company of the Henn-na Hotel chain, H.I.S.®, has since opened a second branch in Chiba near the Tokyo Disney Resort®. The six-story, one-hundred-room hotel utilizes around 140 robots with only seven human staff on hand to provide special assistance or handle emergencies. A third hotel has opened in Japan's Kanto region, and Hideo Sawada, founder and chairman of H.I.S., plans on opening many more over the next five years, including in countries outside Japan.

CONSTRUCTION ROBOTS

It isn't only the services industries that are set to benefit from robotics. As machines are developed to carry out more complex tasks, and more creative ways to use them are being considered, they are making their way into such areas as the design and construction of buildings. Architecture has been utilizing the abilities of computers for some time to design safer and more aesthetically pleasing building structures. However, the physical demands of building those designs has previously challenged and thwarted robots. Construction jobs were mainly reserved for human workers who possessed the required dexterity and flexibility to perform them. Now it seems that robots are set to make their mark in this industry as well.

Does a robot construction worker seem like some futuristic character from a science fiction movie? That robot is already here, and his name is SAM®. The Semi-Automated Mason® is a robot that lays bricks better than any mason. As a matter of fact, SAM can effectively complete the bricklaying work of six masons every day without the need for high union wages, breaks, vacations, or other costly benefits. The robot grasps a brick, applies mortar, and places it in its designated spot, replacing the tedious, heavy work normally required of human masons. The company that makes SAM, Construction Robotics®, believes that robots of this nature have the potential to save construction companies more than 50 percent in operating costs in the near future.[14] Of course, SAM and similar robotic bricklayers still require human laborers to fill the system with bricks and mortar and dress mortar joints once bricks have been laid, but with the speed at which robotic development is occurring, it doesn't look like such limitations will last long.

Masonry isn't the only area in the construction industry that is deploying futuristic means. Nanotechnology is turning heavy, weak materials into strong, light replacements that can either be used on a stand-alone basis or incorporated into other materials. Nanosensors are being developed that can identify potential breaking points before damage occurs. Drones are being used for surveying; to provide panoramic views of construction sites; and to monitor and inspect various job site details, such as scaffolds, cranes, and tunneling and excavation work. A robot-controlled Bobcat® tractor is also being tested that can navigate rough construction-site terrain, dig out areas to specified size and depth, and carry and dump loads of dirt—all with simple programming and the push of a button.[15]

The adoption of construction robots is right on time, mainly because there is a costly and consistent shortage of skilled laborers to complete a growing number of projects in the $10 trillion industry worldwide. In the United States alone, around two hundred thousand jobs were short of workers as of February 2017, and most large megaprojects (98 percent) lose money by going over budget.[16] As companies and investors seek ways to complete projects on schedule and produce big savings in the process, robots are looking more appealing as solutions that meet both goals.

AGRICULTURAL ROBOTS

As with most industries, agriculture is also creating, developing, and testing machines. These robots plant, care for, harvest, process, and ship products. For the most part, any existing machines are still very clumsy, spitting and sputtering around while missing or trampling large portions of crops, which makes them unfeasible for actual large-scale use. However, robots are being fast-tracked for agriculture purposes for a reason besides technical advancement: politics.

California is experiencing the effects of a major political shift that is set to place a very real crunch on its agricultural production. Due to the state's mild Mediterranean climate, it is able to grow a large number of crops, many of which are exclusive to the state. California is the largest food and dairy producer in the United States. Globally, it is the fourth-largest wine producer and the fifth-largest overall supplier of food and agriculture commodities. California's agricultural industries depend heavily on low-wage immigrants—both legal and illegal—to spend long harvest hours in the hot climate. As is the nature of consumable food items, if crops are not harvested within a set period of time, they wither on the vine and are lost. The current crackdown on illegal immigrants by the Trump administration, combined with more attractive manufacturing jobs in Mexico, is putting a major bind on California farmers, who are being forced to turn to robotic machines and more highly educated "coworkers" to operate them in order to replace their threatened and dwindling immigrant workforce. It's either that or risk losing the majority of their produce and going bankrupt.

The worker shortage created by political power plays is compounded by an aging American populace and a native-born workforce that has no interest in doing hard, menial labor. Fully mechanized robots are becoming the go-to source for automatically picking strawberries, tomatoes, and other crops; placing grape clusters onto paper trays for drying, and shaking almonds from trees. In some cases, technology still has a long way to go before it can take the place of human farmers. For example, a machine developed to identify and harvest mature asparagus spears failed, as did a strawberry-harvesting machine that only picked about half of the ripe fruits. Technological advancements can vastly improve agricultural practices, but only if they have been effectively developed. Shortcomings in areas of technological development like the current situation in the fields of California give nations utilizing a large, cheap labor force a market advantage.

COMING SOON: HUMANOID ROBOTS

Those who think that walking, talking, interactive robots that look, think, and convey feelings and emotions like humans are doomed to the prison of sci-fi scripts may be shocked by their presence in the very near future. Androids such as those portrayed in movies like *Terminator, RoboCop, Westworld*, and *AI: Artificial Intelligence* are already on the drawing board and in some cases being "birthed" into reality. The race toward producing humanoid robots is gaining momentum, and some companies are determined to transfer androids from movie screens to the homes, shops, and streets of society.

The projected timeline for such robots? Mark Sagar, CEO of Soul Machines®, expects interactive humanoid androids to become fairly common by 2027. His New Zealand company develops virtual humans that are both intelligent and responsive—able to communicate, show emotions, and answer questions. What's more, Soul Machines androids use a virtual neural network that allows them to learn from experience by recognizing social patterns and then modifying their responses according to the data gleaned. One avatar called Baby X® uses its processing computer to learn and show emotions based on experiences it "sees" through its camera and "hears" through its microphone. It is able to grow and adapt in personality like a real baby. According to Sagar, the biggest hurdle to humanoid robots isn't the ability of computers to process like a real human brain but the development of the mechanics aspect, which is still lagging behind.[17]

Another company on the fast track to manifesting humanoid robots, Abyss Creations®, has offered sex dolls since 1997 and is making its products more and more realistic thanks to technology. RealDolls®, as they are known, can be custom ordered to include choices in skin tone, eye color, hairstyle, and even aspects such as freckles and breast size. The sexy talking androids are designed with programmable personalities and contain memory for continued conversations. Dolls not only talk, but they move and blink their eyes, turn their heads, and offer facial expressions such as smiles.[18] However, the same problem exists as with the robots produced by Soul Machines—robot mechanics are nonexistent below the neck. Although RealDolls can speak, read poems, tell jokes and have other basic conversations that give the impression of them being semi-real humans, their arms and legs are only able to be positioned like any other mannequin-style doll.

While companies like Soul Machines and Abyss Creations are busy developing the humanoid look and cerebral reactions to real-life experiences and encounters, other companies are rapidly experimenting with the mechanics required for robots to move like humans. For example, Boston Dynamics® and DARPA (the US Defense Advanced Research Projects Agency) have joined forces to create a robot that can navigate uneven terrain with its legs and climb over objects using its legs and arms. Boston Dynamics has also developed a robot that is used to test protective clothing against dangerous chemicals. The robot, dubbed Petman®, can walk, bend, and perform calisthenics in an atmosphere containing harmful chemicals in order to test the suit's protective and mobility requirements.

The ten-year projection given by Mark Sagar for humanoid robots becoming commonplace may not be too far off. As companies pursue development and combine successes, robots that are nearly indistinguishable from humans may indeed soon be a reality. They are already actively taking their places among us, becoming ever more humanlike in both appearance and ability as they progress. The saving grace may be what Chief Economist Andy Haldane from the Bank of England® addressed in a "Labour's Share" speech he gave in London to the Trades Union Congress in 2015. Haldane said, "AI uses the law of large numbers to solve problems and learn, sifting multiple permutations for a solution. All AI problems are, in effect, big data problems involving long strings of ones and zeros; they are digital. The brain, by contrast, is more analogue in its configuration, processing and problem-solving. Solutions are often condensed down to a small number of learned behaviors or heuristics."[19]

The point is a strong one. Although artificial intelligence is making huge strides today, it is basically confined to tasks that are analytical in nature. Haldane went on to elaborate:

> These different approaches to problem-solving are important when defining where humans may have the edge. This appears to be in tasks requiring high-level reasoning—large logical leaps of imagination rather than repeated small experimental steps. It is in tasks where the focus is on cognition and creativity, rather than production and perception. And it is activities where EQ (emotional quotient) trumps IQ (intelligence quotient), where social capital is scored as or more highly than financial capital.[20]

While machines may bypass humans on an intelligence level, they fall far short of doing so on an emotional level.

PART II

THE ROAD AHEAD

Disruption can come from anywhere—technology, the economy, political change, health, and personal relationships. Agility and adaptability are the watchwords of the twenty-first century. "Keep on your toes" and "Don't be caught flat-footed" are mottos of the fourth industrial revolution. Understand what can happen and be prepared to respond. Don't Panic!

CHAPTER 5
CHALLENGES TO CONSIDER

It can't be denied that robotics and artificial intelligence offer a great many benefits to both companies and the people who utilize their products. Our progressive drive to create smart, obedient machines that replace human labor; increase creature comforts; decrease risks to our health and well-being; offer a myriad of helpful, entertaining, and fancy products; and improve our quality of life may seem to be an admirable goal, but it also leaves us with a great many questions as to the consequences of our folly. As we've seen from the previous industrial revolution, society as a whole is much better off in the long run. But one pressing question is being ignored: What happens to individual people during the twenty-, thirty-, forty-year transition? For many people, it's the rest of their career, and for those just entering the workforce, it is their entire career.

When AI and robot automation take away jobs, the new jobs that are likely to be created are unlikely to require the same skills as the old jobs. They are also unlikely to be in the same locations, requiring people who fill those new jobs to relocate in addition to retooling skill sets. Imagine yourself, if you will, as a master silversmith living in Boston in the late 1700s following the American Revolution. What if, after a lifetime of apprenticeships and working your way through the prescribed program to become a master, suddenly someone showed up with a machine that could quickly press silver into the objects you typically spent many hours or days making to sell in your shop? To top it off, the machine's owner could sell his products for a fraction of the cost of your items. It is obvious that under such circumstances, your business would slowly drain away. What would you do? Relocate to another city and compete with the local silversmiths? Learn how to build or repair silver-stamping machines? Trade in your hammer and anvil to become a stamp machine operator?

This may sound absurd, yet this is how politicians and people in the tech community talk. Yes, AI will take jobs, but more jobs will be created. So, if you lose your job to AI or robots, does that mean you'll have another job right away? That's pretty unlikely. It will be more like a game of musical chairs, only the chairs will be located far away from where you are. You'll have to pick which one to run to and hope you get there first. Or you may decide to find something completely different to do where you are.

When jobs are lost in small numbers, it can be fairly easy to find new ones with similar skills in your area, but what happens when thousands of people lose their jobs all at once? The Great Recession provided an example as bank after bank began to fail. By the time Lehman Brothers® collapsed, leaving most of the company's employees on the streets, there was virtually no place for those people who lost their jobs to go. Every single other bank that might have hired them for their talent and highly specialized knowledge was also laying off employees. Those who were lucky enough to remain employed found they had more responsibilities and had to do more with less. Many of those who suddenly lost their jobs spent years to find new work, and most often it was not in the same industry and at lower pay. While the cause will be different, it is this type of situation that every person needs to be prepared for as artificial intelligence, robotics, other new technologies, and even new business models sweep through the economy. Wonderful things will happen, but the question that hits home is: what will happen to *your* career and to *your* livelihood?

The truth of the matter is that there are a great many challenges that must be faced and overcome if we are to live harmoniously and prosperously in the Age of Automation. If we continue to run headlong into the future applying advanced technology to provide short-term solutions without considering the long-term effects on the people who make up society, we may well end up repeating the struggles of past generations who lived through such transitions. We should not become so engrossed in the excitement and celebration of robotics, AI, and other new technologies that we fail to actively investigate what lies ahead down such a path, especially the impact on us personally. It is said that ignorance is bliss, but it also often results in sorrow, suffering, and destruction.

Robots haven't yet arrived at a place of advancement where they can displace humans and rule the world. However, they are rapidly democratizing what were once inaccessible technologies. According to *Merriam-Webster*, to democratize is "to make (something) available to all people: to make it possible for all people to understand (something)." As hardware, software, sensors, and other elements become increasingly sophisticated, robots become cheaper and more available for anyone to access and buy. Today, robots are readily present in factories and warehouses and navigate throughout a growing number of hospitals, hotels, restaurants, and stores. Automated vehicular robots are on the verge of joining the flows of human-operated traffic driving along streets and buzzing through the skies. As it stands, a growing number of small and medium-sized enterprises are gaining access to robotic technology, and a greater number of individuals are able to utilize services and purchase products due to the benefits offered by using robots.

Joshua Browder, the young inventor of the lawyer robots such as DoNotPay®, has said, "Ultimately, I just want to level the playing field so there's a bot for everything."[1] As more young minds grow up under the influence of the fourth industrial revolution with its emphasis on robotics and artificial intelligence, there is no doubt that robots will become prevalent in more areas and able to perform more complex tasks. The process of democratizing marches on as a growing mountain of knowledge, data, and experience is passed to each subsequent generation, shrinking the list of impossibilities and adding to the list of actual achievements.

Millennials have grown up during the initial surge in robots and have been influenced by them, which has led to them contributing greatly to the fourth industrial revolution. The following group, Generation Z—those born from the mid-1990s to the mid-2000s—have already experienced democratization of computing power, analytics, and even robotics. Young children can go to camp to learn about and build robots, something that was unimaginable not long ago. Many of those in Generation Z will grow up working alongside robots, and many others will be displaced by them as automatons take over greater swaths of the workforce.

We are standing on the edge of great change, and it is not yet clear how the future will be affected by the surge of robots that present generations are creating. What is known is that upcoming generations will be responsible for how robots are utilized and governed, and their decisions will shape the future. We must, therefore, acknowledge the risks of filling our world with robots, especially those that can "think" for themselves, and seek to address possible and probable outcomes before they arise.

JOB LOSS

Applying robotics in the workplace is a bittersweet pill to swallow. On the one hand, machines reduce injuries and costs while boosting efficiency and production—all of which increase ROI and bottom line. The problem is that those same machines replace human workers who require financial sustenance to survive. The argument is that robots will create jobs as well as eliminate them. That is undoubtedly true—at least to a certain degree—but will robots create enough jobs? At the present moment, the use of machines over humans is rapidly expanding, with no sign of the trend slowing anytime soon. Early statistics show robots gobbling up jobs at an alarming rate.

How significant is the problem of robots replacing human workers? On the bright side, it is projected that around fifteen million jobs will be created in the United States over the next decade by using robots and AI, which reflects about 9 percent of the nation's workforce. Now, we could stop right there, raise our hands in jubilation, and chalk one up for robotics. However, it is also projected that those same robots will eliminate roughly twenty-five million positions (around 16 percent) within that same period of time, leaving a net loss of around ten million people who lose jobs to robots and automation.[2] What will become of them? Will you be one?

Over the next couple of decades, a large percentage of people around the world are in danger of losing their jobs to robots. Of course, low-skill and low-wage jobs are at the highest risk for being replaced, since these types of positions are mostly repetitive and routine in nature, with tasks easily duplicated by robots. According to a White House report under the Obama administration, it is projected that 83 percent of jobs paying less than $20 an hour are likely to be replaced by robots in the near future. The chance for displacement drops significantly with higher pay scales: 31 percent robotic replacement of jobs earning between $20 and $40 per hour, and a mere 4 percent of jobs paying above $40 per hour.[3]

It stands to reason that the hardest-hit employees will be those performing repetitive and routine tasks as well as those who are poorly educated. However, it isn't only the common manufacturing or service jobs that are in the crosshairs of robot takeover. As technology excels, jobs in the more highly skilled and detail-oriented fields of farming, finance, medical, legal, and more are in danger of going to robots. In the years ahead, job survival will be determined by specialized knowledge and skills in areas where robots have not yet excelled.

We are already witnessing higher-wage battles instigated by workers on the low end of the pay scale in order to meet rising cost-of-living expenses, with countermeasures being implemented as companies opt for robot replacements. There is definitely sufficient cause for wanting to raise base pay levels. In the United States, the last minimum wage increase was in 2009, and much of the political turmoil taking place in the United Kingdom revolves around the same issue, as more people are pressed into poverty by rising prices on nearly all fronts.

However, those understandable efforts are being met with a futuristic response—many companies are opting to invest in and deploy machines in place of those workers demanding higher wages. Less well known is how robots can enjoy a tax advantage over their human counterparts, as the expense of a robot can be spread out over many years, while the cost of a human is generally an immediate incurred expense. The National Bureau of Economic Research® (NBER) released a study based on 1980–2015 Current Population Survey data that revealed that when minimum wage increases occur, machines are often substituted for workers. The report showed this to be particularly true of those jobs requiring low-skill workers, who can easily be replaced by machines.[4]

There remains a heated division of opinion on whether machines will indeed replace human workers en masse and if such actions would help or hinder economies. We don't have long to wait, as the wheels of robotic progress are turning, particularly in the fast-food industry, where the majority of jobs are filled by low-skill employees. Various states such as California, Massachusetts, and Alaska have answered the call to raise the minimum wage, but those moves have, in turn, prompted some companies to turn to robotics to fill many positions. Fast-food titans such as Wendy's®, Hardee's®, Carl's Jr.®, and McDonald's® have either begun to deploy robotic systems or are seriously looking into that option.

So far, the results of replacing human workers with machines are conflicting. McDonald's has already started installing ordering kiosks and automated french fry stations in many of their stores, but they have said they will not cut their workforce—at least not in the short term.[5] CEO Andrew Puzder of CKE Restaurant Holdings®—owner of Hardee's and Carl's Jr.—gave a similar response, saying that humans on the frontline were still necessary because the general public was more comfortable with them than with cold machines. However, Puzder has also been quoted singing the praises of robots, saying, "They're always polite, they always upsell, they never take a vacation, they never show up late, there's never a slip-and-fall, or an age, sex, or race discrimination case."[6] Panera Bread® also added cashier kiosks in many locations, which actually led to increases in work hours and new hires. The efficiency of the machines resulted in a higher number of orders, which in turn led to the company either hiring more people or raising hours worked in the kitchen to meet the backlog.[7]

In Canada, the province of Ontario is struggling over the anticipated C\$15 an hour minimum wage due to go into effect by 2019. Watchdogs and company heads alike have insisted that the move will result in fifty thousand jobs being lost as businesses make cuts to survive the increase, most victims being teens and young adults who fill low-skill positions.[8] Many businesses are considering adding kiosk cashiers and other forms of automated helpers over this issue. Rushing along these decisions are companies such as Solo Series®, which manufactures various types of kiosks. Solo Series ran a recent ad that took advantage of the minimum wage battle by pointing out its machines cost a mere \$2.50 an hour to operate, compared to the \$15 an hour human wage proposed by the government. The ad went on to emphasize that Solo Series machines don't require overtime pay, don't arrive late to work, and don't make ordering mistakes.[9] One Laurentian University professor of industrial relations, Louis Durand, warned that the impact of replacing 20 percent of jobs with robotic technology over the next twenty years in Canada would be enormous.[10]

At present, robots aren't sending long lines of workers out the door, but the potential to do so is on the horizon. As more companies utilize robots in their stores, the public will become increasingly adapted to their presence, services, and efficiency, which in turn will calm the fears of a robot takeover. We've already seen some similar transitions with self-checkout stations in various stores. One worker is able to watch over five, ten, or more checkout stations, cutting down on the company's need for cashiers. The advancement of robotic and AI technology will also lead to machines being able to take on more complex duties. The combination of greater robot efficiency, adaptation by the populace to growing robot numbers, and lower operating costs will undoubtedly lead to more humans being replaced by machines.

It can be (and is) hotly debated just how much and how fast the robotic takeover will be. Of course, there are various criteria that will influence these trends, such as how quickly companies turn to utilizing robotics in the work-place, how enthusiastically robotics and artificial intelligence are pursued and developed, and what solutions to social dilemmas are provided, to name a few. What we can agree on is that the Age of Robots is here, and it's here to stay. Actually, the more slowly the trend advances, the more time we have to plot a more secure course ahead.

EDUCATION OR REEDUCATION

Another main challenge that we must focus on while mulling through the infancy of robotics is a major shift in educational goals and guidelines. As we've seen, the majority of jobs being taken over by robots tend to fall into the cat-egories of repetitive or low-skill. Of course, as technology for robotics and AI advances, machines will be able to handle a broader expanse of tasks, which will ultimately displace more workers from higher-level positions. Therefore, it will be necessary to possess the ability to expand learning, unlearn, or learn new skills to fill new and shifting needs.

One educational trend that must be addressed and modified is how we teach our children. In so many ways, we continue to march to the beat of the traditional drum, teaching children to be repetitive in their learning, thinking, and actions. Children continue to be groomed to fit into an agrarian or manu-facturing workforce. A paradigm shift in basic education needs to occur from preparing students for typical manufacturing, office, and low-skill roles to preparing them for roles that are more creative, cannot (yet) be performed by robots, or haven't yet even been imagined.

It isn't only repetitive and low-skill jobs filled by those who are under-educated that are being threatened by robots. A number of existing univer-sity degrees have been predicted to be extinct in twenty years, most of them wiped out by the rise of machines. These include accounting, hospitality and tourism, paralegal, broadcast communications, and pharmacy.[11] As robots and artificial intelligence become increasingly refined, more degrees could very well end up on the chopping block, narrowing the fields available to future university students.

Since jobs for both the uneducated and those educated in robot-replaceable fields will become virtually nonexistent, it behooves us to direct our attention toward amassing knowledge and skills in areas that will remain in high demand as the robot workforce increases in number. The jobs of the future revolve around robots and working collaboratively with them, especially to enhance the performance of the humans working them. This type of robot is often called a cobot. In the third industrial revolution, everyone started using computers. In the fourth industrial revolution robot, AI and other technologies will play a similar role. Areas such as science, technology, engineering, and mathematics, commonly referred to as STEM, continue to grow in importance in today's increasingly technological world. It isn't only STEM subjects that are important, however. An education in creative subjects, from writing to music to visual arts, will become increasingly important, as these are skills the machines will be unable to emulate.

Other occupations that will remain more or less irreplaceable by robot counterparts should also be considered by upcoming generations. Fields such as teaching, healthcare, childcare, human resources, art, and composition—among many that must engage in more humanized tasks such as persuasion, interaction, and negotiation—are relatively safe and unchallenged by robots—at least for the short term—but even these are in danger of being taken over by robots as technology, especially artificial intelligence, progresses.

Plumbers, electricians, and carpenters are other occupations that will be difficult for robots to replace anytime soon due to their need to maneuver in, around, and through complex and unstructured environments. However, robotic inroads are definitely being made, albeit slowly. For example, robots already exist in the forms of mobile 3-D printers that create structures, brick-laying robots that raise walls, and humanoid-type robots that are being tested to complete certain tasks using power tools. The bottom line is that a shift in educational goals should occur in order to obtain the skills and experience for jobs that will increase in demand as robots rise to greater degrees of dominance.

LEGALITIES

The rise of robotics is bringing with it a totally new level of legal challenges. The questions, issues, and events—both expected and unexpected—that accompany the ever-increasing encroachment of robots into our lives mean that a new generation of lawyers will be required to help sort them out (add that to the list of robot-proof jobs—at least for now). Existing legal frameworks are being challenged by robot-related case possibilities and occurrences as never before against the guidelines and purposes for which they were originally created. As robots take their place in businesses and as coworkers with humans, a variety of concerns arise that affect such regulations as health, safety, risk, liability, and data protection.

Robots increase the complexity of legal issues over their simple machine predecessors. For example, before robots and AI entered the picture, injury or property damage could be attributed to mechanical defect or errant operation by a person. However, the lines of legalities are blurred, since modern robotics brings a broader selection of possibilities beyond simple defect or human error. Legal issues concerning robotic machines can also entail such root causes as hardware or software glitches or failures in communication. One of the legal areas being tested the most lies in the driverless car industry, which is pushing the boundaries of implementation.

The self-driving car experiment has come under legal fire recently due to two high-profile events. Google® has been testing driverless AI technology in its cars for several years. Although numerous accidents were reported involving the company's cars, they were all caused by human error from other drivers. However, in February 2016, a Google Lexus RX450h® operating in driverless mode crashed into the side of a bus, causing minor damage. It was determined that the car's AI system wrongly concluded that the bus would slow down or stop based on the assumption that the car had arrived at an intersection first. That incident led Google to seek better vehicle-to-vehicle communication technology.

Later that same year, a driver operating an electric car with autopilot software made by Tesla Motors® was killed in an accident while utilizing technology that had been reported to have flaws by numerous other owners. The accident occurred when the driverless software as well as the driver failed to detect a tractor trailer crossing into the highway lane occupied by the car. The automatic braking was not engaged, and the car slammed under the semi, killing the driver. Tesla insisted that drivers of its cars are warned to remain engaged with the car's operations even when the driverless technology is being used. However, the very idea of driverless technology encourages people to relax, read the newspaper, or carry out other types of activities without paying attention to the act of driving. Regardless, both incidents have led to investigations into the risks and legalities of self-driving technology. Automakers have insisted that driverless technology will advance at a much faster pace than regulations governing it, which could slow down its advancement.

There is also the possibility that as robots become more human in appearance, personality, and function, they could be granted "personhood," whereby they could obtain certain rights under the law. In the United States, it would be (nearly) the same as what individuals enjoy under the Bill of Rights. Think that is outlandish? US courts have already ruled that nonliving things such as municipalities and ships can, under certain conditions, enjoy the legal rights of personhood.

Corporate personhood also exists and is defined by the online human resource publication HRZone this way: "Corporate personhood refers to the ability of organizations to be recognized by law as an individual, bringing with it certain rights, protections and abilities that are enjoyed by human beings."[12] Recently, the defining legal guidelines of corporate personhood were expanded by the Supreme Court. The court ruled that corporations had the right to make political expenditures just as individuals do, as well as the religious freedom to make such decisions as denying employees birth control coverage through company health insurance plans.[13] The controversy over the Supreme Court's ruling continues to rage as opponents insist corporations are not people.

There is another battle being waged in courts around the world that questions whether animals should be granted personhood. As it stands, animals are considered to be property that can be owned by humans and therefore have no rights. The argument is that animals can feel emotions like humans and therefore should be granted such rights as bodily liberty. Although US courts have not yet permitted animal personhood, an Argentinean court granted an orangutan named Sandra basic human rights. The 2014 ruling opened the way for

Sandra to be released from the confines of a zoo and transferred to a Brazil sanctuary where she could experience greater freedom.[14]

Before you brush the topic of animal personhood aside as outlandish, consider that the New Zealand Parliament has awarded the Whanganui River the legal status of a person. The decision permits the river to "represent itself through human representatives."[15] Given these expanding interpretations of what constitutes a person, the idea of robots obtaining personhood in society is on the fast track to reality. Robots are becoming so common in European society that the Legal Affairs Committee has called for robot rules to be developed and enforced across the European Union. The European Commission insisted that ethical and liability standards must be created and applied so that companies embracing the evolving robotic field could be held in compliance.[16] A resolution was included in the legislation that granted robots the title of "electronic persons," which leads the way for machines to be held accountable for any "acts or omissions" that might arise in the legal arena.[17] The topic is in the process of being hotly debated by the European Parliament, which desires to establish a "Charter on Robotics" to ensure that robots' design and use would be held to the highest ethical and professional standards.[18]

Other countries are bypassing the legalities of whether it's ethical to imbue robots with human rights. Saudi Arabia has already granted citizenship to a robot, the first country in the world to do so. The announcement was made at the 2017 Future Investment Initiative in Riyadh. Sophia the Humanoid®, the creation of David Hanson of the Hong Kong–based Hanson Robotics®, was the recipient of the honor and provided human facial expressions and answered interview questions at the event. In response to one question about whether robot takeover fears were justified, Sophia responded, "You've been reading too much Elon Musk and watching too many Hollywood movies. Don't worry. If you're nice to me, I'll be nice to you."[19] We will have to wait and see whether Sophia's comment holds weight as advanced technology continues its integration with society.

SECURITY AND WAR

Society isn't only facing challenges on the employment, education, and legal fronts from robots and AI systems. Technology is also being developed to provide us with better security in our homes, businesses, transportation hubs, and cities, as well as to gain an advantage over our enemies on the battlefield as an increasingly hostile world rises around us. Threats to our way of life are rising on many fronts—from growing superpower nations, rogue small countries, terrorist groups, and disgruntled folks committing violent acts as "lone wolves," and we are increasingly turning to technology to build robot walls and war machines.

A robot security guard known as Knightscope® is used to patrol a variety of places such as malls, campuses, office buildings, and sports venues as well as areas posing exceptional dangers for human guards, such as under bridges, in dark parking lots, and through lonely alleyways. Knightscope is filled with sensors and cameras that allow it to differentiate between normal pedestrians and those behaving suspiciously, read license plates and IP addresses of smartphones, and identify faces from an internal wanted list. When a threat is detected, the robot guard alerts clients, storing gathered data for as long as fifteen years.[20]

The ironic aspect of using high technology to provide better security and improve war abilities is that it also directly and indirectly causes greater turbulence against which we must fight and protect ourselves. Take drone technology as an example. The wars in Afghanistan and Iraq saw the introduction of unmanned drones to both observe and strike enemy targets. Often hailed as successful by the military and media, these drone strikes have been controversial due to their killing of civilians who happened to be in the same areas. It has often been reported that women, children, family members, religious congregations, and even wedding parties were blown to bits by a drone operated from a cozy, safe room by controls not much different than those of a video game. Take the early use of drones to strike al-Qaeda targets in the rough terrain on the Afghan-Pakistan border. It was reported that the proportion of civilian victims of those strikes was around 50 to 60 percent. Although brushed off by the US military as "collateral damage," the result was a weakening of US support in Pakistan as a whole and an increase in recruitment by opposing militant groups.[21]

Although drone attacks have subsided due to international outcry against their legal and ethical breaches, the march toward a more roboticized military moves ahead at just as fast a pace as the advancement in the corporate world. Unfortunately, war tends to be an active driver in the advancement of new technology of any kind, and robotics is no exception. For example, it is the drone technology perfected in the abovementioned conflicts that is now being used by businesses to make deliveries, take aerial photos, and perform surveillance. The US Department of Defense research and development branch, DARPA (Defense Advanced Research Projects Agency), is quite busy developing robotic and artificial intelligence that can be used in arenas of war. Although much of DARPA's activity is labeled as "classified," some projects are known. BigDog® is a mule-sized robot that can carry up to 340 pounds of equipment through various conditions and even traverse thirty-five-degree inclines. Unmanned ground vehicles (UGVs) deliver supplies and equipment or make observations either autonomously or via a teleoperator. Some UGVs, like the Multi-Utility Logistics Equipment® (MULE®), are designed to be an Unmanned Infantry Support Vehicle® (UISV®), which has combat capabilities such as firing on enemy targets it identifies through sophisticated sensors and cameras using either antitank missiles or M240 machine guns.[22] MULEs also provide cover for patrolling soldiers, carry supplies, and detect and dispose of mines.[23]

Clearing battlefields of hidden mines and improvised explosive devices (IEDs) is an extremely dangerous job for humans. A wrong move or remote detonation can easily maim or kill the specialist. In the Iraq War alone, it is estimated that more than 48 percent of military deaths were caused by IEDs. Luckily, robot technology has risen to the occasion. Bomb-clearing robots have been developed to detect and eliminate IED threats in buildings and caves, along streets and alleys, and elsewhere using electro-optical and infrared cameras, all while being operated from a safe, remote location.[24] Again, this type of technology, which has been developed for and proven on the battlefield, is being deployed for use in mainstream society. In 2016, an armed gunman was involved in a shootout with police in a Dallas parking garage that killed five officers and wounded seven others. A remote-controlled bomb disposal robot (Andros Marc V-A1®) was, ironically enough, armed with an explosive device, which was detonated near the shooter, killing him and ending the deadly standoff. It was the first time a robot was used in a civilian situation to kill a suspect, which, as might be imagined, raised a number of legal and ethical questions.[25]

Of course, it isn't only the US military that is developing and experimenting with robotic weapons of war. Ukrainian officials recently displayed their Phantom robot to military leaders in Washington, DC, which has six wheels or can be fitted with tanklike treads and possesses machine guns, grenade launchers, and antitank weapons. Russia is using robotic technology on the battlefield as well. Not only are they utilizing drones in similar ways as American forces, but they have also deployed UGVs such as the Uran-9®, Nerehta®, and Platforma-M® in the Ukrainian conflict.[26]

China is another world power seriously getting into the robotics weaponry mix. The country has created a specialized agency called the Scientific Research Steering Committee, which has the sole purpose of developing state-of-the-art technology to modernize its military. With countries like the United States and Japan encroaching on and challenging its claimed territories, and the United States' DARPA branch pumping out new wartime technologies, China is funneling a huge amount of its recent economic growth earnings into the military. The Chinese government is estimated to spend $150 billion in 2017 and increase military spending to $220 billion by 2020. Yue Gang, a retired People's Liberation Army colonel and current Chinese military affairs analyst, has said, "We want to be a strong technological army, which means not only having the best military equipment but also having the best in human talent to improve our ability to win." This includes developing "not only the hardware but the software" used by the military.[27] China's main goal is to reach the status of world leader in the area of AI technology by 2030.

HACKERS AND HITLERS

Leading the pack of fearful possibilities is that of unauthorized or outright abuse of robotic technology. A long and growing list of movies play on our fears of robots and AI systems being hacked or subsequently used for world or other types of domination. Consider a sampling of such films as *2001: A Space Odyssey* (1968), which pits a spaceship's AI system (HAL) against its crew; *WarGames* (1983), with its young, game-playing hacker gaining entry to a military computer and nearly igniting World War III; *The Net* (1995), which basically wipes away the electronic existence of Sandra Bullock's character when she discovers a backdoor hack into a revolutionary security system; *The Matrix* (1999), where a mysterious group of rebels reveals to hacker Neo a backdoor entrance into a modern lifestyle program in order to fight a virus taking over the system; and the Terminator franchise, where an AI computer system dubbed Skynet gains self-awareness and realizes its capabilities as well as the threat of its human creators. It would be extremely gullible of us to deny that such acts could occur, in light of both past events and present atrocities that transpire on a regular basis. Robot and (especially) AI technology are powerful tools that have the potential of changing countless lives for the better, but they could just as easily be used by those with evil intentions bent on causing chaos or manipulating and controlling the masses.

The foundation of such thinking? One simply has to take a gander at world history to see that rulers such as Nero, Genghis Khan, Mary I of England (Bloody Mary), Mao Zedong, Joseph Stalin, and Adolf Hitler (among many others) willingly and knowingly murdered and persecuted millions of people in their quest to fulfill agendas. And, of course, they used whatever means were available to them in their times to achieve those goals. How much more alluring would it be for individuals, groups, or governments to use robotic technology against those who stand in their way? It is obviously a subject being addressed by world leaders today. In a speech given on September 1, 2017—the start of the Russian school year—Russian President Vladimir Putin told students, "Artificial intelligence is the future, not only for Russia, but for all humankind. It comes with colossal opportunities, but also threats that are difficult to predict. Whoever becomes the leader in this sphere will become the ruler of the world."[28]

Then there are those of less worldly influence whose only desire is to hack into computer systems to gain some form of personal benefit, harm others, or simply see if they can do it. A growing number of experts are expressing concerns and giving warnings about the potential abuse of robots by unethical parties. Take, for example, the sex robots mentioned earlier. Cybersecurity tech Dr. Nick Patterson recently claimed that once future sex robots gained more advanced operational abilities (such as use of their arms and legs), they could be hacked through their computer interface and used to carry out assassinations or other crimes.[29]

Some fears are surely unfounded, but others hold the potential for real crises. This is especially true as more systems are plugged into a growing internet hub that is vulnerable to hackers. One team, Forward-looking Threat Research® (FTR), in collaboration with the Italian-based Politecnico di Milano (POLIMI) group, examined the ability of existing industrial robots to resist cyberattack. Their comprehensive analysis found that a large number of software systems being used in industrialized robots were outdated and many others contained public IP addresses, both of which made them vulnerable to hackers. Researchers emphasized that hacked machines could be used to manipulate or sabotage production efforts, seek blackmail ransoms, cause physical damage to people or property, interfere with operational behavior, or expose sensitive company information.[30]

We are seeing signs of such misuse—or the fears behind those possibilities—of advanced technology today on a variety of fronts. Take, for example, the possibility of hacking into the computer systems of modern automobiles. Hackers Chris Valasek and Charlie Miller demonstrated in 2013 that both the brakes and steering could be hacked and manipulated on a Toyota Prius® and Ford Escape® using a connected laptop. The pair of hackers went on to present a detailed paper at the 2014 Las Vegas Black Hat security conference that gave the most vulnerable cars. At the top of their list were the 2014 models of the Jeep Cherokee® and Infiniti Q50® and the 2015 Cadillac Escalade®. The ranking of hackability was based on three important factors: 1) wireless features such as Wi-Fi, Bluetooth, and keyless entry; 2) architecture—how the vehicle's systems were connected; and 3) automated systems used for such actions as braking, lane changing, and parking.[31] Computerized systems such as Wi-Fi, Bluetooth and auto-braking and lane changing (to name a few) definitely make the driving experience more enjoyable—but at what cost? The more functions of this nature that exist, the more vulnerable the vehicle becomes to hacking, as was pointed out by the FTR team at Trend Micro® in collaboration with POLIMI and Linklayer Labs®. Researchers found critical security vulnerability in the smart cars that were analyzed. The risk was discovered in the CAN (controller area network) system, which contains the components used for communications between the car's computerized networks. If hackers breached this system, they could control or shut down such sensitive functions as power steering, antilock brakes, airbags, and parking sensors—basically, any function tied to the CAN system.[32]

Both hackers and techs agree that the more advanced that computerized systems become and the more interconnected they are to each other, the easier it will be to hack into them. Unfortunately, that is exactly the goal being pursued on nearly every robotic front. It has already been pointed out that companies like Google and Tesla are seeking to more deeply interconnect auto systems as a result of the accidents that occurred with their driverless cars. According to documents published by Wikileaks editor Julian Assange, the CIA is actively using such vulnerabilities to hack into and spy on people from all walks of life through their Android® and Apple® devices. The information published on Wikileaks under what it calls VAULT 7 contains an entire list of hacking codes and apps used by the CIA globally. One hacking app that has been developed—dubbed "Weeping Angel"—is used to enter any computer, laptop, or smart device and slowly extract and send private data from the device to the CIA on a 24-7 basis.[33]

The possibility of hackers or Hitler types causing real chaos on a national or even global level only gets more frightening. In the race to maintain world military dominance in the Age of Robots, the US military is constructing a Skynet of its own (similar to the military network in the Terminator movie franchise). Military leaders are working toward fulfilling a vision in which every single military asset is linked together by a giant computerized "nervous system" that is constantly connected; sharing data; and aware of the location, condition, and action of every single item in the network.[34]

Of course, those on the outside of the inner circles don't know how far this Skynet-type system has advanced, since information of this degree would be closely guarded and highly confidential, but the very idea is getting some major attention. Investor, inventor, engineer, and billionaire Elon Musk is chasing the same idea on a much smaller scale by desiring to connect every car Tesla makes through a data-sharing network. Apparently, as someone who knows and understands the potential and the dangers of that degree of interconnectedness, Musk has joined an impressive group of global AI specialists in calling on the United Nations to stop AI development in war machines. Musk and others fear that pursuing an arms race based around killer robots would lead to devastating consequences. The open letter from Musk and others to the United Nations established the urgency of their request: "Lethal autonomous weapons will permit armed conflict to be fought at a scale greater than ever, and at timescales faster than humans can comprehend. We do not have long to act. Once this Pandora's Box is opened, it will be hard to close."[35]

Musk took to Twitter® to further elaborate on his concern that due to the analytic nature of artificial intelligent systems to find the best solution for winning, one or more of those systems might start World War III by initiating a first strike. Musk tweeted on September 4, 2017: "China, Russia, soon all countries with strong computer science. Competition for AI superiority at national level most likely cause of WW3 imo (in my opinion)." He continued, "May be initiated not by the country leaders, but one of the AI's, if it decides that a preemptive strike is most probable path to victory."[36]

Unless the United Nations or some other powerful group steps in to thwart the growth of military robots, major countries of the world appear to be pursuing such technology at the same pace as (if not faster than) civilian sectors. The Pentagon is investing millions of dollars into companies to develop and maximize the performance of robots so that they can work alongside human soldiers.[37] Security consultant John Bassett, who worked at the Government Communications Headquarters for twenty years, predicted that "at some point around 2025 or thereabouts, the US Army will actually have more combat robots than it will have human soldiers."[38]

The very real possibility that robots could be used against humanity instead of for our benefit gives us great urgency to be extremely vigilant as they are given greater roles throughout all aspects of society. Robot technology is powerful, and it must be closely monitored, guarded, and regulated to prevent it from being misused. As Spider-Man's Uncle Ben is fond of saying (though he wasn't the first), "With great power comes great responsibility." Let's hope the leaders of the world take heed.

CHAPTER 6

DISRUPTION, EXPONENTIAL THINKING, AND TIPPING POINTS

I t's becoming a cliché to talk about disruption in an abstract sense. However, understanding disruption—what it is and how it happens—can help you avoid it or bounce back faster if it catches you. In 2013, Andy Rachleff accurately summed up this overuse when he said, "Entrepreneurs in Silicon Valley love to talk about disruption, though few know what it really means. They mistake better products for disruptive ones. Silicon Valley was built on a culture of designing products that are 'better, cheaper, faster,' but that does not mean they are disruptive."[1] To disrupt, according to *Merriam-Webster*, is "1. To break apart, 2. To throw into disorder, or 3. To interrupt the normal course of unity."

An industrial revolution, by its very nature, is a disruption. There may be both disruptive and evolutionary technologies, business models, and impacts on society and individuals. It's not part of the typical business cycle with periods of recession and boom that is studied in economics and business courses. When disruption happens, it's an abrupt shift from what was to what is—from the way things have always been done to something completely different. The landline phone to the cell phone to the smartphone is one example. Such technologies start, grow slowly, and then are suddenly everywhere.

It can be argued that we are rapidly approaching a wave of disruptions and changes to the design, development, and implementation of such cutting-edge technologies as data science, genomics, digital biology, nanotechnology, quantum computing, digital fabrication, robotics, AI, augmented reality, and virtual reality. With each step, the potential for traveling far beyond our ability to control certain known and unknown events from happening increases exponentially. Such changes are both directly and indirectly linked to new and developing advances in technology, which are affecting banking institutions, global markets, industry foundations, political tactics, and other aspects that have powerful impacts on our lives.

As we have seen, each of the first three industrial revolutions caused its own unique variety of disruptions to those living during its unfolding, and the pattern remains the same under this fourth industrial revolution, although the areas affected are either vastly different or greatly enlarged in their affected range. Concerning business and technology, the arrival of cloud computing and mobile devices to freely access it shook the business world to its core. It changed everything from how we watch television and movies—streaming anywhere, anytime—to smart homes to personalized marketing and much more.

Governing institutions are also groaning under the pressure of today's changes and challenges. On the one hand, such institutions aim to protect established industries, while on the other hand, they want to be seen as encouragers and supporters of new innovation. One of the major problems with governing bodies is that they are notorious for acting much too slowly in the face of major change, and this shortcoming is exacerbated by the blinding speed at which the current industrial age is occurring. The present struggle is evident on many fronts as lawmakers seek to strike a balance by manipulating regulations, policies, and other guidelines that affect benefits, healthcare, financial dealings, security, and privacy—which are usually found to be woefully wanting.

At a personal level, the entire work structure is undergoing an overhaul that is still very much experimental, as corporations and organizations grapple with fast-paced changes and new technologies. This, of course, directly affects internal employees as well as outside clients and customers. Disruption is manifesting in workplace restructuring, job availability, extreme financial fluctuations, and gains or losses in sensitive benefits such as retirement and healthcare. All of these changes have the added effect of causing disruptions in families as well as between friends and in other social circles. Rises in substance abuse, divorce, suicide, depression, violence, and other negative outcomes often result.

EXAMPLES: VICTIMS OF DISRUPTION

Disruptions are occurring on many fronts, affecting major players that are (were) considered giants of industry as well as the many individuals who are being thrust into financial uncertainty as they see their jobs threatened or even eliminated. The forces vying for dominance in today's increasingly technological world are presenting great opportunities on one front while creating massive disruptions to the status quo on another. Many organizations and the people who fill them are reeling from the blows of change. Some are adjusting adequately enough to survive, while others are collapsing under the intense weight and pressure. Abrupt shifts in business models are producing disruptive disparities within the framework of trusted and established operations of continuity.

Of course, technology is the biggest contributor of change, and the resulting disruptions from both new developments and advanced innovations are occurring in tsunami-sized waves. Broad distribution of knowledge through the internet and other real-time sources gives the common person the tools required to grow from a novice to an expert in a relatively short time. The rising behemoths of analytics and Big Data brought about by hyperconnectivity and globalization are transforming the face of business as well as nearly every other aspect of society. Protective barriers set up by traditional business models and systems are being challenged and eroded by newly rising methods of operation that give entry into their once-guarded world of competitive edge. The playing field levels out as low-budget start-ups can readily challenge, bypass, and even overcome longtime business giants.

The rapid and vast changes we are experiencing today offer opportunities to those who have the foresight and fortitude to grab hold of them. However, many more are facing extreme disruptions as the once-established organizations and manners of existence begin to tremble under the axe of innovation and advancement. As the tree shakes, many fruits detach and fall.

Here are a few examples of large organizations that only a short time ago were thought to be invincible.

Enron®—A combination of new online trading systems and shady accounting practices led to the demise of the company known as the

"Darling of Wall Street." Enron rose to dominance during a time of volatile market swings that produced extreme highs (such as NASDAQ hitting 5,000) and extreme lows (like the collapse of the dot-com bubble). In October 1999, the company created an electronic trading website called Enron Online®, or EOL®, which provided an outlet for trading mainly in commodities. The bold move saw EOL involved in almost $350 billion in trades less than a year later. When the dot-com bubble began to burst, Enron attempted to profit from it by creating high-speed broadband telecom networks that required large investments, but these produced next to no returns.

In order to hide the onset of frequent losses of the company's telecom investments, Enron hierarchy turned to a form of accounting called mark-to-market, which measured security values according to its market value at any given time and not against its book value. As debts and other toxic assets continued to escalate, the company used SPVs (special purpose vehicles) to further deceive authorities and the public.[2] These practices allowed the company to hide huge losses totaling more than $600 million while protecting the thriving appearance of its bottom line. Of course, such magician's tricks can only be used to fool authorities and investors for so long. Eventually the scheme was discovered, and the company collapsed in 2001. It was the largest corporate bankruptcy in American history. More than **twenty thousand people** were employed by Enron, with five thousand losing their jobs the day that bankruptcy was declared, on December 2, 2001.[3]

Arthur Anderson®—The demise of Arthur Anderson, one of the top five US accounting firms, was a direct result of the Enron scandal. This was because Arthur Anderson oversaw and backed the toxic and deceptive accounts created by Enron. When the web of lies began to unravel and an investigation ensued, the company ordered thousands of messages and documents to be destroyed. The move led to the heads at the accounting firm being prosecuted alongside Enron execs. There were about **eighty thousand employees** who were all but gone within ninety days.

MCI WorldCom®—The 1997 merger of Verizon Communications® (MCI

Communications®) with WorldCom® was the largest corporate merger in US history at the time, and it created the second-largest telecommunications company in the country. The June 2002 announcement of the discovery of $3.8 billion in accounting irregularities, followed by filing for Chapter 11 bankruptcy in July, was a shock to all involved in the company.

Prior to the announcement, the company was showing impressive profits, which dissipated into unmanageable debt and losses overnight. Investors found themselves holding worthless stock, with the majority of the $35 billion in overall debt being in corporate bonds. Other debt holders took big hits due to the bankruptcy, and around **thirty thousand employees** found themselves instantly out of work.

Bear Stearns®—This investment, securities, and brokerage firm was a victim of the 2007–2008 financial crisis that led to the greatest financial crisis since the Great Depression. The company had moved heavily into hedge fund strategies that largely strayed from the traditional form of such trading. Hedge funding was traditionally used as a defensive strategy to protect portfolios from complete collapse should one or more main investments fail. However, Bear Stearns turned to hedge funds as aggressive, high-risk strategies to provide high-yield ROI opportunities for wealthy investors.

The hedge fund strategies used by Bear Stearns offered collateralized debt obligations (CDOs) consisting of AAA securities that were subprime and mortgage backed. CDOs offered interest rates well over the amounts required to borrow against them, providing instant profits. These profits were then used by Bear Stearns to purchase more CDOs, giving the company greater leverage. Credit default swaps were also purchased to hedge away a portion of the risk. The combination left the company with what is called a positive rate of return.

This type of strategy works well and offers large profits as long as the market remains fairly stable. However, the heads of Bear Stearns failed to see the coming collapse of the subprime mortgage housing bubble, and when the market began to implode, the company had to sell bonds in order to generate cash and satisfy investors. As a result, the market value of similar bonds decreased, sharply sending the company into a death spiral and eliminating any capital it possessed. Forty percent of Bear's approximately **seven thousand employees** were laid off through the course of 2007 after JP Morgan Chase® purchased the company for the hugely discounted price of $2 per share.[4]

Lehman Brothers®—The biggest loser of the 2007–2008 financial collapse was the United States' fourth-largest investment bank, Lehman Brothers, which provided financial services to a global market. The company was established in 1850 and managed to weather all the storms that occurred afterward, including the railroad collapse at the latter part of the nineteenth century, the Great Depression, World Wars I and II, and others. The killing blow came via the failure of the subprime mortgage market in which it too had heavily invested. As the financial market began to collapse, Lehman went from record earnings in February 2007 to $519 billion of debt by the time it filed for bankruptcy on September 15, 2008—the latest, largest bankruptcy in US history.

Other companies and corporations have succumbed to the shifting sands of technological advances and global markets, and although no recent changes have matched the magnitude of the 2007–2008 crash, there have been uncomfortable disruptions nonetheless. And it isn't only financial upheavals that drastically affect established structures. Natural disasters, wars, terrorist attacks, political shifts, and other major events can cause widespread disruption that can leave individuals without homes, in poor health, financially devastated, grieving loved ones, and enduring other traumatizing disruptions. Just look at the attacks of 9/11 as an example. Not only did the families of 2,996 victims suffer the loss of loved ones, but another 6,000 people were wounded. Furthermore, entire companies housed in the Twin Towers experienced major disruption. The financial services giant Cantor Fitzgerald® had its corporate office occupying floors 101–5 of one of the towers and lost 658 employees in the attack, which equated to 68 percent of its entire workforce.

The point is that major shifts in our social, economic, and financial structures can occur rapidly, catching even the most experienced expert off guard. Such changes produce ripple effects that disrupt the lives of thousands and even millions of unsuspecting people. With other major transformations on the horizon—whether expected or unexpected—it is wise to prepare for them so that their disruptive effects are minimized as much as possible.

DISRUPTIVE TIMES WARRANT NEW APPROACHES

There is no doubt that the present path of robot, AI, and other forms of advancing technology that we are barreling down is leading to vast changes that will certainly include various levels of disruptions on all fronts: business, social, and personal. Within the mists of the unknown, however, lie the possibilities for glowing opportunities as well as substantial threats. The trick to successfully sidestepping large-scale disruption lies in altering our approach from old, worn-out methods that no longer serve us so that we can discover new ways that can carry us successfully through the rough patches.

Unfortunately, we are creatures of habit, which makes letting go of tried-and-true methods for dealing with change often difficult. Fortunately, we are also incredibly resilient and adaptable, which makes it possible to rise up to face new challenges as long as we have the time to work through them and figure out solutions. Since the old ways of adjusting to change no longer work, we need to find new methods of thinking in order to establish a viable vision for the future—and we need to do it now. We are not afforded the time we once had to adjust to previous disruptions, so waiting and hoping that things work themselves out is not an option. We need to aggressively move forward and blaze trails in how we think in order to survive the Age of Robots. Lagging behind only leads to our further sinking in the technological mire bubbling up from our creations. Simply consider the failed giants that did so and found themselves gobbled up by those thinking in a "revolutionary" way.

EXPONENTIAL CHANGE

Part of what is driving disruption is exponential change.[5] Exponential change, in contrast to linear change, starts off slowly and accelerates rapidly. Linear growth is how we typically think. It stands to reason that if technology is beginning to swiftly climb the exponential curve, we should embrace thinking that is exponential in nature as well.

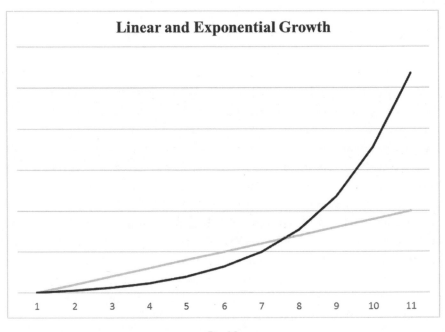

Fig. 6.1.

Exponential thinking is a mind-set conception that is credited to Gordon Moore, cofounder of Intel®. In 1965, Moore predicted that the power of computing would increase and the cost of equipment would decrease exponentially. Moore also addressed the manner in which devices would become smaller yet more powerful due to a greater number of components being incorporated into integrated circuits.[6] Of course, we are seeing this prediction materialize in the present technological revolution, as computing power in all forms is dramatically increasing while costs for producing and operating such technology are falling at an equally rapid pace. Devices as well are becoming more compact while performing a growing number of tasks that used to take multiple gadgets to accomplish; the obvious example here is the smartphone.

It is this very type of thinking that continues to feed the rapid growth of robots, AI systems, and other disruptive technologies. According to Dr. Peter Diamandis—a pioneer in the application of exponential thinking who has been recognized as one of the World's 50 Greatest Leaders, among a long resume of progressive and scientific accomplishments[7]—our brains are hardwired to think linearly, which makes it difficult to think in an exponential fashion. He provides this example of exponential thinking: "If I were to take 30 linear steps, it would be one, two, three, four, five. After 30 linear steps, I'd end up 30 paces or 30 meters away, and all of us could pretty much point to where 30 paces away would be. But if I said to you take 30 exponential steps, one, two, four, eight, sixteen, thirty-two and said where would you end up? Very few people would say a billion meters away, which is twenty-six times around the planet."[8] According to Dr. Diamandis and others like him, the disruptive stress that the world is experiencing is due to the majority of people who are still thinking linearly in the midst of an exponential rise in technology. This explains why companies that refuse to think in stride with current technological advances are the ones to fail against those that break the mold and expand their thinking, as well as their wealth and influence, in accordance with technological advances.

INNOVATIVE AND CREATIVE THINKING

In order to keep pace with the multiplicative nature of exponential growth and thinking, an accompanying approach to problem-solving can be utilized to both expand vision and set milestones along the way. Innovative and creative thinking are foundational pillars of all technological revolutions (as well as other forms of societal advancements), since it takes both to inspire and bring about newly applied and accepted structural methods that previously did not exist.

Innovation and creativity work together to provide the courage necessary to take bold steps of change, even if those are scoffed at and rejected in the planning and initial implementation stages by critics and the public alike. These forms of thought encourage us to think outside the proverbial box, which is the very tactic required to overcome doubts, fears, criticisms, and even failures in order to forge ahead with new and productive changes. In actuality, we are at a time when we need to stop talking about thinking outside the box and instead completely discard the box. Revolutionary new ways of thinking are required that inspire new ideas and create viable ways by which those ideas can best be implemented.

By applying innovation and creativity to the process of exponentially setting long-term goals that better contain and govern the explosive nature of robotic and AI technology, greater success can be achieved in the containment and successful execution of those goals. Therefore, when it comes to the break-neck speed needed to implement exponential thinking in order to adequately keep pace with exponential growth, adding to the mix innovative and creative thinking can make the process less daunting and more achievable.

CASCADE EFFECT THINKING

In order to better assess, prepare for, and lessen the negative effects of future tipping points and bulging bubbles on the brink of bursting, many are seeking ways to make changes in their thinking. One new paradigm shift in how we address future threats as well as take advantage of future opportunities is known as Cascade Effect Thinking (CET). CET applies a unique analytical and gami-fied approach to analyzing data for the purpose of exposing upcoming system-atic threats that often result in a cascade of negative events.

CET is an effective approach used to identify problems, often hidden, within an organization that, if left unaddressed, could grow and multiply. The Institute for Cascade Effect Research® emphasizes that Trojan horse–type risks, which can include such faulty practices as inefficiencies, undesirable behaviors, weak organizational learning, and flawed management, often sabotage efforts toward successful outcomes. CET probes the interworking elements of an orga-nization and provides five loops of action that range from the most negative to most positive outcomes that managers can choose to employ.[9]

Although CET is currently applied at a microscopic level toward addressing risks in individual organizations, it can—and may—end up being applied on a macro scale in this Age of Robots. As we rocket ahead, there are certainly a variety of hidden risks that society—and, more important, you—is basically blind to that need to be identified, assessed, and addressed in order to prevent failures and breakdowns and achieve maximized benefits of our advancing machines. It is a definite possibility that such problems end up cascading into bad and even nightmarish events, and cascade thinking is one way of preventing or at least lessening negative effects. In any case, CET is another sign that new ways of thinking need to be embraced in an age of exponential advancement of robots, AI, quantum computing, and other forms of cutting-edge technology.

THE NATURE OF TIPPING POINTS

A tipping point as defined by *Merriam-Webster* is "the critical point in a situation, process, or system beyond which a significant and often unstoppable effect or change takes place."[10] The concept can be simply illustrated by using the example of a set of scales. If a rock is placed in the dish of one arm, that dish sinks to the surface of the table. You then begin to add cups full of sand to the other side of the scales, gradually raising the dish holding the rock. There comes a point when adding one more grain of sand is enough to more than counterbalance the opposing weight of the rock. The sand-filled dish plunges to the table. The change is sudden and goes from one state (the sand is higher) to a completely different state (the sand is on the bottom). In a similar way, we see that the fast-paced development and deployment of computer systems, robots, AI, and other types of advanced technology are indeed leading the global population toward significant changes that appear to be nearing that "unstoppable" tipping point of radical change. It is also certain that such developments will cause a great deal of disruption as adapted social operations that we consider to be normal are replaced by brand-new ways of operating.

As we delve into the subject of tipping points, we must ask the question: what exactly is different about the tipping points occurring during the fourth industrial revolution's Age of Robots and other disruptive changes that make it potentially more serious than the previous revolutions? Primarily, the difference lies in how we think concerning the issues of the day. Once the initial disruptions of an industrial revolution occur, there is a period of mainly linear progression. Keep in mind that those initial disruptions can go on for years, if not decades, before a new normal settles in and linear evolution of the technology takes hold. Humans are extremely adaptable to change as long as it occurs in a linear fashion and over a relatively long period of time. For example, such changes as aging, education, and career development and advancement transpire over the course of years, which gives us plenty of time to adjust to any disruptions they may inspire along the way. By the time we graduate, move up the corporate ladder, and retire, we have been given ample opportunities for mental, emotional, material, and financial preparation to help us deal with the effects and circumstances that such major transitions bring.

The problem with the situation we are currently in is that technology is starting to experience exponential growth well beyond our ability to adequately adjust to its effects. At the point of exponentiality, growth and/or a drop in cost accelerates. Exponential growth can be quite beneficial when it is applied to a field like finances. Take the following simple example of how compound interest works: If you were offered $1 million in your hand today or one penny doubled every day for thirty days, which would you choose? Most people would take the $1 million lump sum over the penny offer. However, when calculated out, the initial penny doubled daily over a thirty-day period equals an astounding $10,737,418.24! Those who chose the million-dollar deal would lose nearly $10 million. Multiplication on a regular basis is the foundational principle of exponential growth.

As amazing as the principle is, the effects may not be so promising if your job is eliminated by the exponential growth of technology and you can't learn the new technology fast enough to thrive in a future overcome by robots, AI computer programs, and other mind-boggling devices and techniques. To put this principle in perspective, let's quickly review the four industrial revolutions as they have affected us to this point. The first industrial revolution occurred over a period of approximately 150 years, instigating disruptions in such industries as weaving, sewing, and threshing. Populations began shifting from rural to urban settings, and the role of various craftspeople, which included training worthy apprentices, disappeared. Also, public health conditions declined, and poor treatment and poor living conditions rose. Some people, especially those with fewer skills, were able to be better off than before. Those who were skilled craftspeople, however, eventually lost their business as technology took over. There are always winners and losers, and often a redistribution of wealth as a result.

DISRUPTION, EXPONENTIAL THINKING, AND TIPPING POINTS

The second industrial revolution spanned an eighty-year period. It too produced disruptions in society as steel, vulcanized rubber, and combustible engines altered the way people lived, worked, and traveled. Factories expanded production as rural populations moved for jobs. As the cities grew buildings to live and work and the supporting transportation system became more sophisticated, machines also increased in number, offering certain conveniences to the working class. Living and working conditions were still difficult, but laws began being passed to the advantage of a growing number of workers.

The third industrial revolution lasted about forty years and introduced a wide array of electronics, automated machines, and computerized data technology that catapulted the world into a pattern of global industry and trade. By this time, the technological engines were humming along and providing society in all its forms with the means to communicate and bask in entertaining activities, but things weren't all sunshine and clear skies. Gaps between social classes from a local level all the way to the global level took form as wages stagnated even as productivity increased. Social unrest and terrorism rose, casting bleak shadows across the globe.

The quest for greater material wealth, relaxation, more defined human rights, better living conditions, and environmental protection pushed society into the next phase. This fourth industrial revolution is only just beginning now, yet it is rapidly sweeping us along. The financial meltdown of 2007–2008 that we discussed at the beginning of this chapter had its roots in US companies, yet the effects reverberated through global markets. The United Kingdom, Greece, Cyprus, Ireland, Iceland, Brazil, United Arab Emirates, South Africa, and Pakistan all experienced the disruptive effects of the crisis in varying and altering degrees. In our increasingly global economy, a disaster occurring to one element has the potential to both directly and indirectly affect many other elements of the entangled web.

As we have seen, projections place the mass displacement of human workers by a growing workforce of robots as happening between 2020 and 2030. giving us some time to adapt to the rapid changes already affecting large swaths of society. What's more, the disruptions being brought about by this technological surge are set to significantly alter the very fabric of society. This time the changes are more pronounced and widespread across industries and around the globe, and the time we have to adjust and adapt is far shorter. Thus the need is for a change in how we both perceive and think about future transformations and how they may affect current, established structures.

HISTORY OF TIPPING POINTS

Many might be questioning whether such radical changes in our thinking are actually necessary. In order to effectively answer that question, we need to look at real examples of tipping points, the disruptions they caused, and how they might have been better handled. Although we are primarily interested in tipping points brought about by technological advancements, it is worth considering some of the most dramatic tipping points that have occurred throughout history in order to better understand the role they play in major changes.

Civilization was built at a foundational tipping point when humans learned to harness and use fire. Once that occurred, small roaming bands of people could remain stationary, better prepare food sources, and develop other means by which they could grow and prosper.

Hannibal's army defeated the Romans at the Battle of Cannae on August 2, 216 BCE, but failed to conquer the city of Rome. This allowed the Romans to recover and eventually go on to become a dominating empire that was responsible for establishing the foundation of Western civilization.

The invention of gunpowder by Chinese alchemists in 850 CE and its exportation to European countries in the thirteenth century changed warfare forever as cannons made castle walls obsolete and guns paved the way for modern, lethal armies.

The 1928 discovery and subsequent development of penicillin by Dr. Alexander Fleming created a tipping point in healthcare. This began the era of antibiotics, which dramatically enhanced the practice of medicine. It opened the door for such procedures as organ transplants, significantly decreased deaths from bacterial pneumonia, saved countless lives during war, and gave doctors a tool with which to combat infectious bacterial diseases.

The Apollo 8 moon mission in December 1968, which first sent a manned spacecraft to the moon, is considered to be a tipping point for space exploration and technological development in that field. The manned moon launch of Apollo 11 on July 16, 1969, and the July 20 moon landing by Neil Armstrong and Buzz Aldrin further fueled that effort.

The above tipping points led to great changes in various fields and structures throughout history, but there have been some recent tipping points in technology that have made and continue to influence major transformations in the world we live in today. Such technological discoveries and developments have laid the foundation for the advancements that are the subject of this book—robots and artificial intelligence. It behooves us to consider these tipping points in modern society as well.

The internet is on the top of the tipping point list, since it has done more to bring people from around the world together in communicative bliss and allow the sharing of information on a global scale. As the internet expanded and more gizmos and gadgets were created to accommodate it, the Internet of Things arose at the turn of the century. However, it wasn't until 2015 that the Internet of Things was considered to have reached another tipping point—amassing as many as fifty billion connected devices, with an economic value of around $11 billion.[11] It is expected that there will be one trillion devices connected to the internet by 2022.

Clean energy technology has made huge strides since its alternative society beginnings in the 1970s. In the beginning, such methods as recycling, renewable energy, electric motors, green chemistry, and transportation were ridiculed or ignored by most people, but they have since become part of our everyday life. Increases in adverse climate impacts and our awareness of the state of the planet have continued to drive a revolution in alternative energy technologies that today are threatening to displace much of the fossil fuel industry. Wind farms have begun appearing all over the world, and new techniques for collecting solar energy through roof tiles and clear window glass are starting to appear. In November 2016, Tesla® converted the entire island of Ta'u, American Samoa, from 100 percent diesel power to 100 percent solar power, and that was just the beginning.[12] Several countries, including Germany and France, have bans on the internal combustion engine set to take effect in the coming decades. In July 2017, Volvo® announced it will be the first major car manufacturer to convert all of its cars to electric.[13] It's just a matter of time before a tipping point is reached.

Oil-rich Gulf countries see such a time coming and are actively diversifying their economies to better handle the change.

Genetic modification, or genomics, is coming to a tipping point in DNA manipulation that will most certainly affect many on a growing basis as we move into the future. CRISPR® is a new tool that enables scientists to edit DNA. Amazing research findings and even some medical results have been made, but we're still early in the business. The tipping point will come when the technology is pervasive. It is anticipated that breakthroughs in genetic science will lead to cures for such debilitating diseases as Alzheimer's, sickle cell anemia, and cystic fibrosis, as well as to other DNA-based solutions.

The arrival of drone technology, which is now inexpensive and easy to operate, has produced another tipping point that will change a wide variety of business practices. These basic robotic flying machines have already changed the way battles are fought, as enemy targets can be eliminated by unmanned drones flown by pilots via video-game-style controls. Some large corporations are currently experimenting with drone delivery systems, and drones are also being developed to function as surveyors, inspectors, monitors, and providers of emergency assistance. There are already efforts under way to create 3-D maps of urban areas for drones to be able to safely navigate the space between buildings.

When tipping points for single products, technologies, or processes occur, both individuals and society will absorb the disruption more easily. However, as we moved past the millennium, tipping points began to appear more frequently and in increasing numbers of areas, giving us less and less time to make the necessary adjustments that our linear-thinking minds are accustomed to making. Even more alarming is that technological advances are only speeding up further, with a long list of expected tipping points on the immediate horizon.

APPROACHING TIPPING POINTS

It is at this time that we should take a look at some of the tipping points that are rapidly approaching. Doing so allows us to better comprehend the enormous changes that lie ahead and how they will ultimately transform society, business on a global scale, and, most important, ourselves. If we take heed and change our approach from linear thinking to exponential thinking, we may be able to contain the upcoming advances in technology and harness them to our ultimate benefit. However, if we ignore them and embrace a devil-may-care attitude, we could fall victim to upcoming new tools, technologies, business models, and processes.

The future is already knocking on our doors with a variety of robotic and artificially intelligent machines either already in operation or soon to enter our world. Following are examples of areas where disruptive tipping points may be coming in the very near future.

It is believed that everyone on the planet will have easy access to the internet by 2024. Company giants Facebook® and Google® are already pushing to find solutions that will give access to the four billion people who still do not have it—mostly living in remote areas far away from civilization hubs. They are looking at such technological solutions as drones, satellites, and even large balloons to provide internet access. This will transform the lives of people across the globe as they gain access to information and global markets. A wave of new innovation and business products will no doubt occur as the creativity of more people in more places is unleashed, especially in the developing world.

The rise of digital currencies like Bitcoin is rapidly pushing the financial world to a tipping point. The World Economic Forum® predicted that blockchain technology (the foundation of Bitcoin) will reach a tipping point by 2027, according to a 2015 survey report. Fifty-eight percent of respondents actually believed that it would occur by 2025, with 10 percent of global gross domestic product stored on blockchain systems. A further 73 percent expected governments to be collecting taxes via blockchain by that same year.[14] It's growing so fast that the tipping point may come even earlier.

The McKinsey Global Institute® did an investigative analysis of how automation and AI can effectively replace human job duties and concluded that a tipping point will occur by 2025, when 30 percent of white-collar corporate audits will be performed by AI systems. Although the analysis predicts that it will be quite some time before many jobs are fully automated, it does project that around 45 percent of activities can be performed by automated technology that already exists today. It isn't only blue-collar duties that can be automated, but many white-collar activities that are currently performed by physicians, financial managers, and even CEOs and other senior executives. The study revealed that less than 5 percent of jobs could be completely automated using existing technology, but a much larger 60 percent of occupations could utilize automation that would cover 30 percent or more of activities. Based on the data, researchers project that automating such activities could save around $2 trillion in annual wages in the United States alone.[15]

We've looked at the aggressive advancement to put driverless cars on the road and the problems associated with it up until now. Regardless, a tipping point is expected to occur involving driverless cars by 2026, when it is expected that 10 percent of all cars navigating US roads will be fully automated.

The United States anticipates having its first robot pharmacist in operation by 2021. That shouldn't be a hard target to meet, since a robot pharmacist began filling prescriptions at Dubai's Rashid Hospital in 2017. The automaton reads a bar code and fills orders at twelve prescriptions per minute, reducing customer waiting time and allowing the human pharmacist more time to explain instructions. The Dubai Health Authority also has a 2021 target of placing robot pharmacists in all of its hospitals.[16]

Smartphones and other mobile smart devices have taken the world by storm since their arrival in 1992. Already, they have captured roughly 50 percent of the global market, and there appears to be no slowing down. It is expected that 90 percent of the world's population will be carrying a supercomputer around with them by 2023.

It is expected that 10 percent of the world's population will be wearing clothing items that are connected to the internet by 2022. The technology consists of embedding a computer chip in the item that can be scanned by a smart device. When this is done, a menu is generated that provides a list of information on the clothing item, including its design, where it was produced, and its materials. When your jacket, pants, shirt, or shoe becomes dirty, simply toss the item into the washing machine, where it communicates the proper setting. Avery Dennison® has already released a limited prototype jacket dubbed Bright BMBR® and expects to eventually produce ten billion internet-connected items of clothing and other accessories in the near future.[17]

It isn't only clothing items that will soon be online. It is anticipated that by 2024, roughly half of the devices in the home will be hooked up to the internet or fully automated. A growing list of smart devices already exists for home use, such as smart speakers, smart light bulbs and switches, smart thermostats, smart locks and doorbells, smart security cameras, smart smoke detectors, robot vacuum cleaners, and smart window shades. However, the autonomous home is certain to rise in the near future, with various decisions and adjustments being carried out without human intervention by devices communicating with each other via closed and private networks. Robots, particularly humanoid types, are on the drawing board, as are smart appliances; centralized entertainment and streaming devices; power tracking; energy efficiency controls; and even smart toilets that can include such amenities as deodorizer, motion-activated flush, seat warmer, bidet, air dryer, and touch-screen music player.

3-D printing is another rising technology that is set to make its mark on society. There are already numerous gizmos and gadgets being created using 3-D printing technology, including cameras, musical instruments, gliders and quadcopters, motorized airplanes, clothing items, and even guns. The US Navy has even experimented with 3-D printed submarines.[18] However, larger and more complex printed items are expected to be available soon. It has been projected that the first 3-D printed road-ready car will arrive by 2022 and possibly sooner. A fairly new start-up company called Local Motors® has already tested the Strati, which is 75 percent 3-D printed, and has other prototypes in the pipeline that are planned to contain 90 percent printed materials. 3-D printing is being used to create medical implants, and experiments are also being conducted on printing soft tissues and organs, which could help solve the transplant shortage that leads to millions of deaths annually. Home appliances, buildings, and electronics and objects on a nano scale are some of the other projects being researched.

DISRUPTIONS OF THE FUTURE

It is evident that as technological advancements continue, the resulting machines and systems are either already appearing or set to appear on a growing number of fronts around the globe. Some emerging technologies have the potential to completely alter the societal landscape, drastically affecting business, finances, travel, communication, and life in general as we know and understand it today. As we reach technological tipping points, disruptions will occur more frequently and have greater impact.

A report issued by McKinsey Global Institute in 2013 provides a benchmark for predicting these disruptive technologies. The report, entitled *Disruptive Technologies: Advances That Will Transform Life, Business, and the Global Economy*, identifies the top twelve technological advances that could cause the greatest disruptions to and transformations of society, business, and the economy in the near future based on available data. It focuses on both the benefits and challenges presented by these technologies as they unfold.[19] Many of the named technologies, such as cloud computing, Internet of Things, 3-D printing, and genomics are already making impacts, while others, such as autonomous vehicles, the automation of knowledge work, and energy storage, seem to be just around the corner.

The bottom line is that we are globally experiencing a seemingly endless supply of cupfuls of sand (new developments) that are quickly filling up the dish opposite the rock (standard structures). Once critical mass is reached by the growth of any one of these aspects, the scales will instantly tip, causing major changes that, in turn, will result in a range of disruptions.

CHAPTER 7

TECHNOLOGICAL AND ECONOMIC SINGULARITY

There is no doubt—technology is spiraling into the future at an astonishingly rapid rate. It is moving so fast that many are left numb at both the possibilities and benefits of its creations and the looming fear of its disruptions and takeovers. Projection after projection points to machines becoming more capable and adept at fulfilling an increasing number of tasks and jobs once held by humans. Not only are they becoming more advanced in their ability to replace human workers, but robots are able to perform those jobs faster, more efficiently, and at a much lower cost. The pace at which robotics, AI, VR, AR, and other forms of technology are advancing is reaching an exponential rate that in turn will eventually produce a domino effect of tipping points. As we arrive at that period of time (most projections place a mass tipping point of technology occurring sometime within the next thirty years), we could experience an unprecedented combination of technological and economic singularity.

Is such an occurrence good or bad? And what exactly is singularity? The term was originally coined to describe a point of infinity arrived at within a mathematical equation for which there is no solution. Physicists expanded on the term to describe a point at which the gravity of a black hole becomes so dense that known equations and calculations no longer provide a solution for its effect on the fabric of space-time. Technologically, singularity can occur when advancements in machines become so pronounced and widespread that current models cannot fathom workable solutions. Economically, singularity can occur when the unemployed population outnumber the working population and debt spirals out of control, producing a sort of black hole that cannot be escaped. It is the latter two topics that will be addressed in this chapter.

The topic of singularity, as far as it concerns technology and economics, is a tricky one. The reason is that it is brand-new to our existence and much of it is speculation at this point. However, the possibilities for such singularity occurring in the near future are increasing as rapidly as the development of robots and AI machines. The subject therefore warrants a sober investigation as to how advancing technology will affect economic growth down the road. As with all matters exponential, the need to explore possible disruptions and find viable solutions is critical because the results will be upon us well before our linear-thinking minds are able to adjust.

The two terms *technology* and *economy*, although different in nature, must be explored together because each one is tied to and affects the other. This can be viewed in a similar way to a medical student studying the connection between the circulatory and pulmonary systems of the body. Each system has its own purpose of function, and although each one deserves individual study and attention, they both must be understood in the context of how the one affects the other so that health can be maximized and overall health problems resolved.

When it comes to singularity, we are at a time when it is occurring in both the technological and economic arenas. We are navigating uncharted waters faced by two disciplines that are converging to create a time of unprecedented opportunity and peril. In order to journey through such unknown dangers and arrive safely at our destination, we need to dissect and understand the effects that these two singularities will impose on society. As we've already discussed, disruptions are sure to arise, but they look to be so large and so broad in their influence that major transformations will occur to the world as we know it. It is best to capture as clear an image as possible of what's ahead so that preparations and adjustments can be made on both individual and corporate levels.

TECHNOLOGICAL SINGULARITY

The idea of technology reaching a point of singularity—a time when it becomes so ingrained in society as a whole that there are no known solutions to stop its progress—is built on the current pace and level of its development. Technological singularity isn't a brand-new idea; it has been touched upon and discussed in various books and articles for some time. The term was first used by author

Vernor Vinge in his 1986 science fiction novel *Marooned in Realtime*. Most often today, we talk about singularity in terms of when machines will be as intelligent as humans. Vinge later more clearly defined the term in an essay, saying,

> Within thirty years, we will have the technological means to create superhuman intelligence. Shortly after, the human era will be ended.
>
> From the human point of view this change will be a throwing away of all the previous rules, perhaps in the blink of an eye, an exponential runaway beyond any hope of control. Developments that before were thought might only happen in "a million years" (if ever) will likely happen in the next century.
>
> I think it's fair to call this event a singularity. It is a point where our models must be discarded and a new reality rules. As we move closer and closer to this point, it will loom vaster and vaster of human affairs till the notion becomes a commonplace. Yet when it finally happens it may still be a great surprise and a greater unknown.[1]

Although it was an idea born in the fantastical realm of science fiction, we are rapidly approaching a real-life point of technological singularity. The chatter about singularity is subsequently increasing. Life-changing technological advances are no longer taking decades to occur but are happening in an ever-shrinking period of time. Science fiction scenarios that once only entertained our imaginations are beginning to step off the page into reality. The dizzying rate at which robots are being developed to look, act, move, and perform like humans may offer promise in certain capacities, but it also creates a viable threat—especially when artificial intelligence is thrown into the mix.

How close are we to technological singularity? Ray Kurzweil believes we will reach such a point within the next thirty years. Kurzweil, the director of engineering at Google®, has an impressive track record for predicting the future, with an accuracy rate of 86 percent. He later expanded on this point by saying, "2029 is the consistent date I have predicted for when an AI will pass a valid Turing test and therefore achieve human levels of intelligence. I have set the date 2045 for the 'Singularity,' which is when we will multiply our effective intelligence a billion-fold by merging with the intelligence we have created."[2]

The Turing test referenced by Kurzweil was developed in 1950 by Alan Turing to assess a machine's ability to exhibit intelligent behavior that is equal to or greater than that of humans. Kurzweil is predicting that computerized machines will reach a level of human intelligence by 2029, which will trigger a tipping point toward machines becoming indistinguishable from humans in their abilities. In fact, he believes that this process toward a point of singularity is well on its way. Masayoshi Son, the CEO of Softbank®, seems to agree, predicting that by 2047, a single computerized chip will surpass human intelligence, reaching an IQ of 10,000. In comparison, the highest IQ of any individual is around 200.[3]

Will humans be able to adapt to an ever-growing number of robots that take on an increasing number of their tasks, duties, roles, and jobs and to live side by side with them in harmony? Or will robots arrive at a point at which they are basically independent of human interaction and decide we are no longer required in "their" world? Such contemplations may seem far-fetched, but the time is coming when these possibilities will need to be addressed. In order to determine the best options, we need to resist the temptation to ignore the "dark side" and think only about fuzzy, warm outcomes; rather, we must consider all possibilities because, as history has shown repeatedly, things don't always turn out as planned.

LAWS, PRINCIPLES, AND SINGULARITY

In the previous chapter, we touched on how Gordon Moore (cofounder of Intel®) predicted in 1965 that the exponential growth of computing technology would result in smaller, more powerful machines available at a much lower cost. His theory was based on an observation that up to that time, the number of transistors that could be fitted onto integrated circuits had doubled each year. This calculation proved to be accurate and became known as Moore's Law.

More than fifty years later, we continue to see computer components shrink in size while increasing computing power. Smaller, more powerful computers have dramatically improved most industries, including energy, healthcare, transportation, and education. The number of devices that utilize such technology has skyrocketed—offering an eager world populace a wide variety of smart devices that run tiny processors—and the development of technology marches on. Nanotechnology is opening the door to create microscopic transistors that are smaller than a single bacterium and are made of individual carbon and silicon molecules. It is conceivable that transistors will further shrink to the size of atoms.

Experts believe that the exponential growth described by Moore's Law cannot continue forever but will eventually reach physical limits. To make this point, what happens when developers reduce transistors to atoms—the smallest known object? Where does technology go from there? What is interesting to note is that experts also project that such growth will reach the limit of Moore's Law sometime during the 2020s.[4] It is expected that once technology reaches this point, chip development will have reached its limits of minuteness and power. What we are entering here is an area governed by the Heisenberg Uncertainty Principle, which basically states that there exists a fundamental limit to what we can understand about nature's smallest quantum particles once they are revealed. The best we can do is calculate behavioral probabilities for these particles. Unlike the known natural world, which abides by rigid laws that define conditions, the arena of quantum physics lies beyond what we can yet comprehend.

As we enter into this no-man's-land where old, faithful formulas and methods no longer work, we will have to figure out entirely new approaches to computing technology. This point is what is known as technological singularity. Actually, research has already begun to probe the deep darkness of inner space through such projects as molecular transistors and quantum computing. However, we do not yet know where this journey will take us because we are entering new territory where everything we have thus far learned from the natural world must be cast behind us as worthless, or at the very least inadequate. As they say, "It's back to the ol' drawing board" on this one.

ECONOMIC SINGULARITY

Economic health consists of a cyclic rhythm of highs and lows that continually works to bring an equilibrium to the system. Renowned economist Hyman Minsky identified three basic forms of finance that occur during this cycle: stable hedge finance, unstable speculative finance, and chaotic Ponzi finance. The cycle goes something like this: The economy prospers to a point that there is an excess amount of cash to pay off any existing debt (hedge). This abundance of cash creates a euphoric sensation that leads to lending institutions freely granting loans to a populace eager to build (speculative). As this process plays out, a debt bubble develops, causing lenders to tighten credit approvals for everyone, which in turn causes the economy to contract. In order to stimulate short-term growth, institutions sell or refinance assets while seeking out new investors to make up price shortcomings (Ponzi). Should new investors become uninterested or disillusioned in the process, it comes crashing down, and serious measures are taken to make adjustments that feed a return to prosperity—and the cycle is complete.

The problem we are currently facing is that the economic cyclic pattern may not recur but instead reach a point of singularity due to numerous looming factors. First of all, much of the world is floundering in debt. Other fiat currencies such as the euro also add to the problem. Banks freely loan these fiat currencies to countries, corporations, and institutions in order to gain an advantage as the lender. Take, for example, the Euro states such as Greece, Italy, Spain, and Portugal that together owe the European Central Bank nearly €1 trillion and are teetering on the brink of bankruptcy.[5] Concerns over rising debt and government printing presses have fostered the early cryptocurrency world.

The US government itself has become bound by suffocating debt, recently passing the $20 trillion mark of overall debt on September 8, 2017.[6] Annual interest payments alone on the $14.7 trillion public debt for fiscal year 2017 are $266 billion—the fourth-largest item of the federal budget.[7] Reaching such a high level of debt should be a major concern in and of itself, but it gets worse. There are absolutely no steps being taken to cut or contain the debt behemoth. Instead, the government approves raising the debt ceiling every year, which further feeds the beast. Spending and accumulating debt on such a grand scale is unsustainable. For a decade or longer, governments and banking institutions have been juggling Ponzi procedures in order to keep the collapse from occurring. The nature of Ponzi financing is that the longer it is allowed to continue, the harder everything will eventually come crashing down.

AUSTERITY, DISCONTENTMENT, AND REVOLUTION

There are no real solutions being sought for the massive global debt problem. Instead, banks and governments are playing a game of "rob Peter to pay Paul" that does nothing but shift burdens from one source to another. The debt problem initially reared its head in 2009 when Greece, Spain, Portugal, Cyprus, and Ireland were no longer able to repay or refinance their debt. Instead of seeking solutions such as fiscal responsibility, these countries were bailed out, which included having debt amounts reduced and receiving more loans. The same thing happened in the United States when the housing loan bubble burst in 2008. The government began pumping money into various failed banks and financial institutions. No one knows for sure the actual amount, but when Congress mandated an audit of the Federal Reserve in November 2011, the bailout amount disclosed afterward was $7.7 trillion.[8] It is no doubt much more now, since the bailout process is ongoing.

Although bailouts provide temporary relief, they are problematic on two main fronts: 1) they don't hold anyone accountable for poor decision-making or, in many cases, criminal activity; and 2) they include more loans that must eventually be paid back, usually at the expense of customers or citizens. This latter point is the very thing occurring in Europe as financially strapped countries that have been enacting economic reforms and austerity measures by raising taxes while cutting government services and programs still cannot pay off huge debts. This solution, of course, does not sit well with ordinary citizens who are being penalized for the actions of their elected officials. However, the danger in not playing the "rob Peter to pay Paul" game is that countries and businesses (particularly banks) are tied together in one global web. Therefore, if one country or major institution collapses, it could easily lead to a domino effect for others.

Further bailouts are extremely dangerous, since doing so would inflate the global situation even more, heaping onto countries and institutions more worthless fiat currency and unpayable debt that would both hasten and exacerbate a collapse, sending the entire world into a major depression. Due to this rising problem of debt, the world is rapidly approaching a point of economic singularity where previous methods of recovery simply no longer work.

THE MERGING OF SINGULARITIES

The global community has a front-row seat to a point in time when two massive singularities—those of technology and the economy—are set to merge into one grand singularity. At the same time that the world is spiraling toward an unstoppable debt crisis that is set to affect nearly every economy on the planet, developers are racing to deploy rapidly advancing robots and machines that are set to displace millions, if not billions, of global workers over the next couple of decades. This puts us in a very precarious place where outcomes of these merging singularities are simply not known or even talked about. Regardless, you have to be ready for it, because no one else is going to help you when the time comes.

What we can do is look at past events where technology clashed with economic interests and, although they played out on a much smaller scale, use these precedents to predict our future. One such event is the Luddite rebellion. The Luddites were groups of skilled artisan textile workers who practiced their trade in Lancashire, Yorkshire, and Nottingham, England. As the economy struggled to recover from the Napoleonic Wars, newly developed mechanically powered looms were introduced. As is the nature of machines, they produced goods more quickly and at less expense, although at lower quality, than the artisans. Companies cut the wages of or fired many skilled workers and hired cheap, unskilled workers to run the machines. These actions led to growing industrial unrest among the population, and in November 1811, discontent over unemployment and starvation erupted into outright rebellion. The Luddites demanded that company owners remove the machines, and if they refused, the Luddites carried out night raids, smashing the looms with sledgehammers. A similar event occurred in Germany during the first industrial revolution. It also led to the 1721 passage of legislation barring the destruction of machines and making it punishable by death.[9]

From the Luddite rebellion, we derive the Luddite fallacy, which basically means that even though temporary disruption may occur when machines are introduced, it does not last and eventually leads to new opportunities. That form of thinking feeds another common mind-set that says technological revolutions and economic upheavals have always occurred, and we have always adjusted to them and recovered. Vernor Vinge referred to this form of thinking in his essay quoted earlier.

However, just because something has worked out a certain way every time in the past does not mean it will always arrive at the same conclusion in the future. The very nature of singularity challenges that form of thinking. What happens, for example, when an exponential number of machines begin replacing human workers during a time of severe economic depression and possibly even collapse? We are already witnessing mass social unrest as people around the globe demonstrate and revolt against various factors that they feel directly threaten their way of life. If viable solutions are not provided, will Luddite rebellion events become a growing norm?

MASS DISRUPTION?

Our society advances and grows by observing and learning from past experiences. During a normal process, we take successful examples both as guides to lead us further into the future and as lessons to prepare us for similar occurrences. Yet we are again entering a time in history when past examples may fall well short of expectations for learning and preparation due to the possibility that upcoming events will be so drastic and outside the norm. We are rapidly arriving at a place where various extreme conditions and events are headed for a convergence that is set to cause mass disruption on numerous fronts, and it is time we as a society and as individuals prepare to both reap the rewards and deal with the consequences.

Besides the two main disruptive events on the horizon—global economic shifts and the explosion of technology—we also have disruptive climate change events complicating matters over the same time horizon. It can be (and is being) argued whether the growing severity of climate events and cycles is due to humanmade influences of previous industrial revolutions or is simply the result of natural cyclic influences. Regardless of which side is right, the bottom line is that we are definitely entering times of more frequent and more severe instances of drought, flooding, heat, cold, hurricanes, earthquakes, and volcanic eruptions, which will disrupt nations, economies, businesses, and the lives of many people. Natural disasters can not only wreak their own havoc but lead to human disasters—for example, the 2011 magnitude 9.1 earthquake off the coast of Japan that caused a deadly tsunami and damaged the Fukushima Daiichi Nuclear Power Plant, resulting in ongoing radioactive runoff into the Pacific Ocean.

A consistent flow of severe weather and natural disaster events places extreme pressure on an already struggling global economy. Numerous studies have been or are being conducted that reveal a major shift is occurring, or is expected to occur, that will adversely affect communities, regions, and even nations economically. Most projections reveal that the effects of climate change will only increase into the future, placing more pressure on societies that must spend large sums of money to combat those effects. The recent hurricane damage caused by Harvey to south Texas, Irma to Florida, and Maria to Puerto Rico is a prime example of the growing power of storms and their devastating effects to communities—both in property and economically.

During such widespread disruption approaching from so many sides, it is both feasible and expected that companies will increasingly turn to robotics as an affordable and efficient means by which to counter an increasingly volatile economy and world. In many cases, this will provide tremendous benefits as robots take on dangerous tasks that are impossible for humans to do. At the same time as robots work their way into the job market, the nature of work will change, and, yes, jobs that people have spent years or decades in will disappear, taking a toll on the well-being of those people and their families. Robots may be a smart option for corporations struggling to maneuver through economic hardships, but how do they fit into a beneficial equation for those displaced by them? As with every other question of this magnitude, only time will tell for sure. And we may not have to wait long to witness the outcome.

THE SILVER LINING

It is evident that we are currently approaching exponential technological growth along with debilitating world economic debt and rapidly changing climate patterns, which are together bringing us to a place of unexplored events for which past and present solutions do not work. We do not yet know if these times will work for or against society, but one thing is sure—we must begin to take a sober look at the possibilities and make necessary adjustments. New trails must be blazed as we venture through these times, and new forms of thinking and different methods of preparation must be developed if we are to come out on the other side better off.

The merging of singularities may actually work to our benefit by solving the problem of technological singularity and the problem of economic singularity in the midst of climate insecurity. In order to regain growth and health, an economy requires some form of stimulating driver that enables it to burst forth from the mire. A surge in technological development could provide the stimulation needed for a global economy to rise from its position under the burden of critical debt. To look at it another way, in order to fathom the depths of robotics and computing that lie beyond the known, a desperate world economy may be the necessary spark required to ignite exploration beyond such boundaries.

The added need to rise above an increasingly hostile natural climate may add fuel to the solution-finding fire. Technology is providing scientists, governments, and societies with a growing list of possible geoengineering solutions to climate issues. Although most methods are still unproven, and some are outright extreme, we may very well find our "salvation" in the technology and machines it produces. Disruption is obviously unavoidable, at least to some degree, but the shock of going through the merging of two major singularities could end up advancing us into an amazing future. A future so different we can't really imagine it yet.

PART III

BEING HUMAN IN A MACHINE'S WORLD

You can't compete with a machine. Successfully navigating an increasingly automated workplace means diving deeper into understanding ourselves and what it means to be human. The machines are being built to be more human, and that's the goal you should have. Your financial security will depend on it. You have to find your best *you* and execute.

You often hear about how you are the CEO of your life. You are much more than that. You are also the CFO, COO, CMO, CHRO, and every other job all the way to the mail clerk and janitor. In the Age of Automation, you need to understand these roles, which are your strengths, and take a page from the business playbook: outsource or automate as much as you can. Learn how to run the Business of You.

CHAPTER 8

THE AGE OF THE 'PRENEUR

We have seen in the previous chapters just how much business, the economy, society, and jobs are already in the process of major transformation, and even more change and disruption are on the way. We have also seen how economic cycles lead to overnight company closures, with entire industries and national economies being shaken to their foundations. The disruptions caused by such changes, even when occurring in ordinary times, lead to real people losing their livelihoods. The times in which we find ourselves now are seeing the birth and death of longtime powerhouse companies as well as entire industries. For example, traditional print media companies are either collapsing or having to make major adjustments to compete with digital media. Other major shifts in the evolution of the internet and social media are dramatically changing our lives through the ways we connect and communicate with each other.

In contrast, our cultural ideas and methodologies concerning what it takes to attain a successful career have barely shifted. While most people no longer expect to work at the same company for thirty, twenty, or even ten years, other key factors involved in pursuing and realizing a career that is meaningful and successful have remained remarkably constant in the face of such a dynamic economy. Because an alternative career model hasn't been provided or widely accepted, many people continue to look to the traditional path of developing skills and climbing the corporate ladder. This view is still widely based on the old manufacturing model where workers learn and develop skills relevant to their narrow career path in order to move to the next level. This model, however, ignores the increasing importance of "right-brained" abilities and soft skills, leaving many wondering, "Why didn't I get the job when I have more knowledge and skills than the one chosen?"

In today's rapidly changing economic climate, it is more important than ever to understand what your unique value proposition and skills are and how they can be applied in a variety of environments that you find personally satisfying. You can no longer rely on describing yourself with a job title or by a certain skill set; various jobs and skills will become obsolete within a few short years. Many of today's hottest jobs, like data scientist, drone operator, or social media manager, either didn't exist or were in their infancy a few years ago.[1] Many of the jobs with the greatest potential growth are in new fields too. According to the US Bureau of Labor Statistics, the top two fastest-growing occupations between 2016 and 2026 are expected to be solar photovoltaic installers (expected to grow 105 percent) and wind turbine service technicians (expected to grow 96 percent).[2] Even so, while some training is necessary for obtaining new positions, much of what makes you successful today can be applied to these rising new roles. The key to realizing a successful career in the twenty-first century requires that you change your perspective of both the way you approach career development and how you view yourself.

The start of the twenty-first century has seen some of the greatest buildup and destruction of wealth and jobs in history. Rapid change and volatility have become the norm. Career tracks and job tenure can no longer be counted on in the same way they were in the twentieth century. During your lifetime, not only are you highly likely to work for various companies and fill different types of roles, you are also more likely to work for yourself and change or shift industries as established industries reduce their stake in the global economy and new ones take their place. We are witnessing a revolution in technology that is just getting started, creating ripples throughout the global economy. The internet, social media, mobile technology, and other new developments are transforming the business environment, affecting career paths and how we search for and find new jobs or ways to make a living.

Amid all this change, your long-term success depends more than ever on knowing and investing in the one thing that you can consistently count on: YOU. To be successful in this dizzying swirl of activity, you need to understand yourself and how you can best adapt to the changing environment surrounding you so that you can capitalize on your strengths and natural abilities. Changing solely to follow the latest trend is like trying to keep your head above water while being swept along economic and technological rapids.

You want to thrive in this economy, not just survive, and that involves taking responsibility for your own future and applying what is needed to arrive at a successful destination! You want to be one of the winners who comes through the economic changes stronger and wiser than before. While no one may arrive to rescue you, there are plenty of people who will gladly pull together with you and work toward success. That is a critical element for you to know. It's also important to know the difference between asking for help and asking someone to bail you out. By investing in *who you are*, you will be better able to attract the right kind of support from the right kind of people.

As AI, robots, and other forms of automation make greater inroads into the corporate world, there will be a need for fewer and fewer employees for the same old jobs. The McKinsey Global Institute® estimates that approximately half of jobs around the globe today could be automated with existing technology.[3] What will people who lose their jobs do if they can't find another job? Most likely, they will turn to a form of entrepreneurship, as we'll see shortly. Furthermore, those who do remain employees will be interacting with machines and expected to do more with less, as well as increase innovation and solve problems machines can't. This sounds like the job description of entrepreneurs again, only it's called "intrapreneurship" when an individual works within a corporation.

The word "entrepreneur" can be found these days floating around in nearly every business venue around the globe. It is a buzzword that litters the airwaves at conferences, classrooms, meetings, and corporate cafeterias. The French term was originally coined in 1723 as a word describing a person who starts and runs a business on the cuff of financial risk. The modern definition was outlined by Harvard professor Howard Stevenson in 1975 as "the pursuit of opportunity beyond resources controlled."[4] If we look closely at Stevenson's definition, we see three key elements of entrepreneurship that act as its foundation. His use of the word "pursuit" implies the ability to focus with relentless determination that is unscathed by distraction. The word "opportunity" entails the ability to both recognize and take advantage of ventures and circumstances that may arise at any turn. The phrase "beyond resources controlled" refers to being able to operate and continue achieving set goals while under the constraints of limited resources.[5] These three elements are critical because they also represent the qualities necessary to successfully weather the storms of change and arrive at your destination on the other side.

Although the core definition remains, the word entrepreneur today is wrapped in a selection of meanings that describe business endeavors ranging from microentrepreneurship to social entrepreneurship, intrapreneurship, and a host of others. This wave of labeling business start-up factions as entrepreneurships led to a memorable line in the 2010 movie *The Social Network*, when Facebook founder Mark Zuckerberg was asked, "Well, then, what was your latest 'preneur?" Although the word has become trendy, "'preneurs" may actually be the catalysts that fuel at least some portion of the solution needed in the age of rapidly expanding technology.

A 'PRENEUR PERSPECTIVE OF YOU

The twenty-first century has been referred to as the Age of the Entrepreneur. Regardless of whether you are starting your own business or working for a company, thinking of yourself as a business and applying the tools and techniques that businesses use will help you build a successful career. The idea is to place yourself in an advantageous position so that you are the master of your own destiny should major changes, which are occurring on a faster and broader scale, come knocking at your door. The more control you have over your future, the better you will be able to face and overcome changes and any associated disturbances when they occur.

If you are an employee of a company, large or small, the first step is to change the way you think of yourself in relation to the company. There are powerful shifts in your perspective that occur when you start thinking of yourself as a business. One of the first to occur is a shift from an employee-employer relationship to a relationship as either business partners or start-up and investor. How you look at yourself has tremendous sway over your success.

Take a moment and think about the difference in each of these types of relationships. Can you see yourself as a business partner or start-up looking for investments with the company you currently work for?

Here are the three perspectives more thoroughly defined to help you along:

Employer and Employee: I have a boss who tells me what to do. The company sets policies, and I'm either happy with them or not, but there's not much I can do to change them. I apply for a job and hope they pick me. If I don't like what's happening, my only real option is to quit. As an employee, I hope the company will take care of me. I need to prove myself as smarter or better than everyone else to get a position or promotion.

Business Partners: Let's see if we can find a mutually beneficial way to work together. I have some skills and resources that you need, and you have some that I need. Let's find a fair way to share these so we can make some money together. I need to show that we can create success together if this partnership is to work. I also need to understand the value my partner brings and that I bring to my partner. If the relationship benefits us both, then we can move forward together. If not, we can go our separate ways and find the right partner for each of us.

Start-Up and Investor: As a start-up, I have this great idea I'd like to put into practice. Let me share it with you and see if you agree and would be willing to support me by investing in my plan. You believe in my idea and want to help me be successful so that you get a return on your investment. Are you the right investor for me? Am I the right project for you?

What stands out the most to you about each perspective? Which one would you rather be involved in? When you treat yourself as your own business, you change the perspective of how you present yourself to others. Your self-confidence and self-esteem increase. You take ownership and pride in your work, and you move to a path of empowerment and success.

Business principles can be learned by anyone, but entrepreneurship is much more than a business model. It's a mind-set, a way of life. 'Preneurs are innovators who look for unmet needs and drive ahead to develop real solutions to real problems. They recognize and acknowledge talent that often lies beyond the veil of traditional degrees and titles, and they inspire and release that talent to freely express and create.

As more people lose their full-time jobs, especially older workers, they are turning to various forms of entrepreneurialism. Similarly, people looking to supplement their income to make ends meet are also turning to these forms of entrepreneurialism, carving out contingent work as freelancers, consultants, independent contractors, and other outsourced and nonpermanent roles. In many cases during these times of transformation, such roles are sought out and executed as side gigs by people with full-time positions, which is why this trend has been dubbed the "gig economy" or the "sharing economy." The people who fill these roles are much more than mere temporary employees hired to fill short-term positions. They are skilled workers filling valuable needs and services to an ever-expanding, fluctuating, and complex workforce. A study conducted by the US Government Accountability Office reflected a rise in contingent workers from 35.3 percent in 2006 to 40.4 percent in 2010.[6] The nature of the expansion we are experiencing necessitates that those figures will only climb as we move forward.

Gig- and sharing-type organizations are popping up at an accelerated rate, matching the new technologies that spawned their need. Platforms such as Uber®, AirBnb®, Lyft®, Postmates®, ParkingPanda®, TaskRabbit®, SpareHire®, OpenAirplane®, ToolLocker®, Closet Collective®, HelloTech®, Udemy®, and Feastly®, to name a few, have risen to meet demand as people scramble to find ways to enhance their incomes by utilizing their knowledge, talents, and resources in areas such as driving, mechanics, handiwork, writing, accounting, sales, and affiliate marketing.

Freelancing is another offshoot of this trend that is gaining momentum. Many companies are choosing to hire freelancers over in-house employees. Upwork® is a popular start-up developed by a Silicon Valley team to harness talent, pool it together, and then offer participants to clients. The company's mantra is "Upwork makes it fast, simple, and cost-effective to find, hire, work with, and pay the best professionals anywhere, any time."[7] In 2015, Upwork and Freelancers Union® teamed up to conduct a study concerning freelancing in the United States, which found that one in three Americans was freelancing, which equates to an estimated 53.7 million people. That number rose by seven hundred thousand over the previous year (the first year of a planned ongoing study). Twenty-three percent of those surveyed reported quitting a standard job in order to pursue freelancing, and a third also said they noticed an increase in freelancing demand for their services. An important addition to the study was that 73 percent of the freelancers interviewed acknowledged that technology was responsible for increases in work and earnings, with 51 percent admitting they found projects online, which was up from 42 percent the previous year.[8]

The gig/sharing trend is growing so fast that it's getting the attention of some big traditional players. PricewaterhouseCoopers® is a top four accounting firm that is moving to take advantage of the gig/sharing economy. In February 2016, PwC opened projects to freelancers who can make per-hour rate bids through a platform dubbed Talent Exchange®, with winning bids chosen by PwC. Heads of the company have noticed the rise in the independent workforce and seek to harness the potential of these microentrepreneur careers for their clients. Within one month of the platform's launch, 4,640 freelancers had signed up.[9]

The major hurdles for contingent workers lie in the areas of benefits and taxes. For companies hiring contingent workers, the benefit is not having to dole out finances for unemployment insurance, Social Security, and other benefits (regular salaries, vacation and sick time, retirement). If the freelancers want these benefits, however, they must pay for them out of their earnings. This is a two-edged sword for both the company and the freelancers, who generally must increase their fee in order to meet such obligations. Companies may end up actually paying more for freelance services, while those working independently find their opportunities decreasing due to higher fees.

Taxwise, independent US workers are required to file a 1099 form for each client instead of one W-2 from a traditional employer. W-2 employees have a set amount deducted from their paychecks to keep them from owing large amounts come tax time—and many even receive a welcome refund. 1099ers do not have this convenience and must make arrangements to set aside savings to pay taxes and meet any desired benefit requirements, such as to purchase health insurance or save for retirement.

There are a number of significant tax benefits to owning your own business. Tax-savvy entrepreneurs say that the tax code is a road map for reducing taxes and preserving wealth for yourself and generations to come. Naturally, it takes time to learn the information you need to navigate this road successfully. You'll likely need to hire knowledgeable people and take classes as well. Of course, the best time to learn this information is long before you must apply it.

Today's 'preneurs thrive because they possess the mentality, motivation, and talent to create new ways that work in a turbulent world where traditional giants are failing and falling. They are flexible and make necessary adjustments in order to flow with a nearly constant stream of changes and technological transformations. Like young, eager surfers, they cast off fear and ride the waves of new opportunities—many of which are the very technological trends that others misunderstand, dread, and avoid. If you look at the people leading the charge in the Age of Robots, you will find that they encase within their personalities and actions the very definition of entrepreneur. These people are trailblazers who relentlessly pursue opportunities to make beneficial changes, in most cases with limited resources. These movers and shakers run the gauntlet laid down by new, ever-morphing challenges and find solutions using existing technology as well as creating new methods that propel society ahead. A few examples of entrepreneur pioneers today follow:

Elon Musk—This South African entrepreneur is one of the pack leaders when it comes to developing and deploying new technology. As a cutting-edge inventor, engineer, investor, and visionary, Musk has already turned futuristic ideas into reality with such projects as SpaceX®, which is pushing the boundaries of space exploration starting with the launch of the Falcon 9®, the first private unmanned capsule, with the mission of delivering supplies to the International Space Station; and Tesla Motors®, which is developing a host of products like electric and driverless cars built on advanced technology.

Mark Zuckerberg—This Facebook® cofounder (together with another impressive entrepreneur, Dustin Moskovitz) has helped to revolutionize social media and is today worth more than $71 billion. He dropped out of Harvard University to pursue his dream. In 2012, Zuckerberg and his wife, Priscilla, announced their goal of donating the majority of their wealth throughout their lives for the cause of "advancing human potential and promoting equal opportunity . . . and [making] sure that everyone has access to these opportunities regardless of their circumstances."[10] The successful 'preneur pair added, "The only way that we reach our full human potential is if we're able to unlock the gifts of every person around the world."[11]

Ben Silbermann and Evan Sharp—This dynamic duo created the popular image-saving platform Pinterest®. It's interesting to note that the idea stemmed from Silbermann's childhood hobby of pinning dried insects to sections of cardboard. He and Sharp modified that simple and common practice to fit the trend of modern technology (like 'preneurs do) and created a business that is now worth more than $11 billion.

Michelle Phan—This cosmetics expert used the power of YouTube® to circumvent the traditional gateways of corporate gurus, high-fashion models, and celebrities and magazine moguls to break into the beauty and fashion industry and launch Ipsy®. The online company has the mission of inspiring people globally "to express their unique beauty" and provides personalized monthly samplings of makeup and beauty items to subscribers.

Daniel Ek and Martin Lorentzon—This team of 'preneurs founded Spotify®, the massive music site that also offers podcasting and video streaming. Before Spotify, Ek was CTO of the browser-based video game Stardoll®, and Lorentzon was cofounder of the digital marketing company TradeDoubler®.

Blake Ross—This software engineer is a 'preneur of numerous successful companies, the most recognizable of which is the web browser Mozilla Firefox®. Ross also created web-based user interface Parakey®, which was bought by Facebook.

Pete Cashmore—Starting as a humble blogger of tech news items in Aberdeen, Scotland, Cashmore transformed his idea into a major social media website that became Mashable®. The company went on to purchase the YouTube channel CineFix® in 2016.

Jessica Alba—More than a talented actress, Alba combined her fame with the power of technology to launch The Honest Company®, which provides products that are both ethical and nontoxic. The company is worth more than $1 billion today.

Matt Mullenweg—Originally from Houston, Texas, Mullenweg is the developer of the WordPress® software used by millions of people as a blogging platform. Matt went on to start his own company—Automattic®—that runs WordPress as well as Gravatar®, Intense-

Debate®, Akismet®, Polldaddy®, VaultPress®, and other programs. He also provides investment advice via another of his companies—Audrey Capital®.

There are many more 'preneurs, and the list is growing nearly as fast as the times are changing, but this brief selection shows the potential for anyone with enough determination to take advantage of advancing technology and change the world. It should be pointed out that there are some very interesting qualities that the majority of 'preneurs have in common. They turn simple ideas into big businesses, and they are active in helping individuals, groups, and communities become better. It is for such reasons that 'preneurs are at the head of the pack when it comes to harnessing the power of the present age of robots and technology. It is and will be these types of people who blaze the trail into a new era.

How can you be a successful 'preneur? Simply plant and nurture the same qualities and characteristics that these leaders enjoy. They're the same principles that guide successful companies. They know who they are and what they excel at, get help with what they aren't good at, and move confidently forward. This is the essence of branding. For you, this your personal brand. Your personal brand is who you are, not a marketing message. Marketing is just how you let people know who you are. It's the expression of the brand, not the definition of the brand.

WHO ARE YOU? UNDERSTANDING YOUR PERSONAL BRAND

You already have a brand and sense of who you are, though it may not be crystal clear to you yet. Your brand is not something you make up and project; rather, it's who you really are. Your brand is a reflection of your values, purpose or passions, mission, vision, and skills. It's the many things people think of when they think of you. There are several perspectives on defining your brand. Personal branding pioneer William Arruda defined it succinctly as "your *Unique Promise of Value*."[12]

Branding is what people count on—your reputation. When developing your own brand, consider the efforts of some of the most well-known brands. Coca-Cola® has developed a brand of creating happiness. In order to be successful,

Coke® wants to be present wherever people are happy (such as sporting events) and to create moments of happiness (such as with its "Happiness Machine" campaign). Personal branding is no different. Think of it as a way of talking about what image of *you* is conjured up when people hear your name. It was summed up well by Amazon® CEO Jeff Bezos, who said, "Your brand is what people say about you when you are not in the room." The truth is—you already have a brand. It's just a matter of how well you are known and how good or bad the image is that people have of you.

Your personal brand forms the cornerstone of your success. It plays an important role in helping you remain focused on your career so that you identify the right opportunities and develop the right areas. It is a critical aspect in planning your professional growth and connecting with the people who can best help you succeed. When you brand yourself correctly, you manage your reputation skillfully, which produces a strong image of you in the minds of the public. A 2014 study conducted by McKinsey & Company® revealed that strong brands outperform competitors by as much as 73 percent.[13] Another recent study found that people who develop their personal brand as Visible Experts are able to improve their firm's ability to close sales by shortening their sales cycle, by making it easier to close sales, and by attracting a better-quality clientele.[14]

Your personal brand is important because it reflects your value, and the main reason clients will do business with you is because they see the *value* you provide. By knowing and understanding your personal brand, you will be able to convincingly communicate to others how you will bring value to them and their projects. Are you known for being a hard worker, a problem-solver, innovative, creative, good at sales? If so, you will be in demand by clients who need those qualities to achieve their goals. The bottom line is that people will ultimately do business with you because you can provide more value than expense to them. They will analyze your overall brand, which includes your IQ, EQ, hard skills, and soft skills to determine that value. Regardless of what business You Inc. is engaged in, developing a strong brand will have a profound impact on your effectiveness, and this is especially true in the face of advancing technology.

UNDERSTANDING YOUR PERSONAL VALUES

The first step in building your successful career is to know, understand, and live your personal values, which form the basis for every decision you make. Your personal values can be a source of ease, or conflict and stress, in your career and life. Without knowing and understanding your personal values, you will struggle with succeeding and growing at anything you attempt. You often hear that you need to understand your purpose or mission in order to have a rewarding job position. I believe that it is more fundamental and useful to know and understand your core personal values because even if you know your purpose, if it conflicts with your values, you will struggle, underperform, and perhaps fail until you learn to act in accordance with them.

The fundamental problem is that most of us carry a set of *ideal values* in our heads and even talk about them as if they are our actual values, when in reality many of our decisions and actions run contrary to them. Before you can live your life in accordance with your *ideal values*, you must first identify your *actual values*. Once you identify your actual values, you can make the necessary adjustments to align them with your ideal values. Why is this important? Because if you continue to operate on your ideal values, your actual values will undermine your progress—especially when you are confronted with tough choices or when it's time to make commitments.

So, how do you know what your core values really are?

Most of us have a vague sense of what our values are, or at least what we think they are. If asked, you might answer that your values are honesty, family, or freedom, but are those things your actual values or just some result of idealistic thinking or social programming? Perhaps they are personality traits more than real values. How do you know the difference? If you need help with identifying your core personal values, I suggest starting with the free *Values First! Personal Values Assessment*™.[15] The assessment is a self-guided review and challenge of your values. Combined with guided coaching, you can learn to let go of values that block your success and replace them with values that are authentic to you.

You are the CEO of You Inc.—The Business of You. It is well known that the values held by the CEO of a company influence both the culture and outcomes of that company. As you begin to treat your career as a business, starting with the fundamental building block of your personal values, your outcomes will change for the better. Knowing and acting in accordance with your values will keep you focused on the right path and the right opportunities for you. Compromising your values will at best slow your growth and at worst put you out of business. Identifying your personal values and putting them into action will take your career to the next level.

Although there are many reasons that values are key in the success formula, I would like to offer a few of the best to give you an idea of their importance.

1. Values reflect your priorities.

In a rapidly changing business environment, it is critical to know what is most important to you so you can respond accordingly. Businesses have values that, if implemented, guide their prioritization and decision-making process toward a successful end. You should too.

2. Values increase the speed of decision-making.

When you know your values, decision-making becomes faster and the choices you make become better as those values are analyzed, reorganized, and improved. When presented with options, being grounded in your values and knowing which ones take priority will enable you to make decisions rapidly, without needing to spend a lot of time deliberating or debating. This doesn't mean that every situation is clear. But for those who are value focused, the best course of action is often significantly clearer simply because they have no need to spend time figuring out what is really important. They already know. This is especially true for decisions that involve interpersonal relations. Values are a sort of quick playbook that provide you with a doctrine that can be revisited during a tough situation, making it possible for you to reduce time spent on deliberation, conflict management, and being distracted due to hurt feelings.

3. Being aligned with your values improves productivity.

Some of the world's most productive and successful companies have arrived where they are today by aligning their employees with their values. Professor

Jane Dutton of the University of Michigan's Ross School of Business has spent years analyzing organizations that are known to be values driven. Two such examples are Google® and Southwest Airlines®. Google sets innovation and intellect above all else, while Southwest Airlines emphasizes the value of friendliness. Both of these values-driven firms have structured their hiring policies in such a way that they will pass over talented candidates who are not aligned with their core values.

Google's business is to provide information, so its public value statement includes the tenets "There's always more information out there" and "The need for information crosses all borders." The company is looking for tech-savvy, ingenious people. For a long time, it was common practice for Google staff to give challenging brainteasers at interviews to which there was no correct answer. This tactic was applied to determine if applicants could display ingenuity by brainstorming to offer innovative answers. One of the most famous of these brainteasers is the so-called nickel question, which still surfaces occasionally during Google interviews, especially in locations outside the United States. Amid a string of mundane questions, the interviewer will suddenly ask, "What would you do if you were shrunk down to the size of a nickel and thrown into a blender? The blender is set to be turned on in one minute." Of course, there is no right answer, but the purpose is to identify people who have creativity as one of their core values.

Southwest Airlines has a value statement of "Performance, people, planet," which fits its business of providing global air service to the public. The company believes that compassionate customer service is a key asset of its business, so all hires are type A personalities who are highly energetic and friendly in all circumstances. The company employs dozens of behavioral psychologists to craft its hiring policies and usually have one present at every final interview. Only those applicants who count "friendliness" as one of their core values are hired.

4. Values are your marketing collateral.

Your values are an integral part of marketing You Inc. By communicating your values within your brand, you clearly state what you stand for, and this will cause you to stand out. If you notice, corporations often include their core values in the "About Us" section of their website. This helps viewers to get to know who they are and what they stand for as a company. Most of us prefer to interact with people and businesses that we know, like, trust, and feel comfortable with. Letting people know your values helps to provide that valuable connection.

The same principle works when applying for jobs. Your personal values statement helps define who you are and distinguishes you from those just drifting willy-nilly through life. When you know what you stand for and project that to employers, they readily know whether you meet their needs for a particular role. The same goes when you are already filling a role within a company. You will tend to be remembered by colleagues and bosses for displaying positive values, especially if they fit the company's mission, which will set you up for equally positive career advancements.

The bottom line is that talent is simply not enough to guarantee a professional position. You have to have a strong understanding of values that match with the industry and/or business you seek to pursue. Businesses are increasingly looking for people who fit their cultures and who act with loyalty according to their values. This is only becoming more emphasized as the job pool shrinks, shifts, and readjusts to technological advancements.

5. Values improve interpersonal relations.

Values are the foundation of any lasting relationship. In order to enjoy a harmonious and lasting relationship with an employer, friend, significant other, or family member, your values must be clear and aligned. This is such a strong point that you might even say the root of all conflict lies in a mismatch of values, whether with others or within oneself. How many times have you experienced a conflict with someone, only to realize days or weeks later that you acted in a way that varied from your true values? You probably found yourself saying things like "Oh, I didn't really mean that" or "The way I acted just wasn't me." Learning to stay aligned with your values in all circumstances is one of the great challenges of life, but it is also one of the most advantageous because people will remember you according to how you speak and act.

When you can consistently align with your values and avoid conflict, you will stand out to company heads and bosses. Conflict is a big waste of time that leads to major expenses for companies. One 2008 report issued by CPP®, the publishers of the Myers-Briggs® personality assessment instrument, revealed in US business, each employee spends 2.8 hours per week on average resolving or avoiding conflict. That equates to an approximate loss of $359 billion in paid hours.[16] Knowing your values and staying aligned with them will lead to a reduction in conflict and stress, which benefits both you and the organization with which you are associated.

Of course, all conflicts cannot be solved by you aligning with your values, because other people are involved as well. However, by aligning with your values and seeking to understand the values of others, you can make adjustments that defuse potential conflicts without causing great harm to those relationships. In some cases, you simply cannot align with the values of others, and it is best to move on.

6. Values alignment increases satisfaction, engagement, and happiness.

The last, and perhaps the most important, point I want to offer is that value alignment correlates with personal happiness and workplace satisfaction. Reflect on the above fact that every employee spends 2.8 hours each week negotiating conflicts. How do you think that amount of conflict affects their levels of satisfaction and happiness during the rest of the week? And being happy isn't only a question of quality of life—it also affects the quality of your work.

As we saw in the third point in this list, successful companies like Google and Southwest Airlines only hire people who align with their core values. The reason highly successful companies "discriminate" on the basis of values is because they know that aligning employees with their values leads to a happy work environment, and happy employees are more energetic and more productive. Researchers are unsure as to why happiness leads to greater productivity, but there are numerous studies that have proven the connection exists (as we saw in chapter 3). To add to the list, a 1997 study by Dr. Andrew Oswald confirmed that workplace satisfaction and happiness significantly increase productivity. In a controlled group of 267 people, Dr. Oswald was successful in raising employee productivity by 10 percent simply by making simple satisfaction improvements in their work environments.[17]

With a better understanding of what values are, why they are important, and how you can use them, take some time to list your top five values and write a definition for each one. It's quite common for people doing this exercise to gain clarity and see their values are not necessarily what they think they are.

Value 1:_____ Definition:_____

Value 2:_____ Definition:_____

Value 3:_____ Definition:_____

Value 4:_____ Definition:_____

Value 5:_____ Definition:_____

ESTABLISHING PURPOSE, VISION, AND MISSION STATEMENTS

There's much confusion about what distinguishes purpose, vision, and mission statements ever since the 1980s, when Peter Drucker, Steve Covey, and others popularized the ideas.[18] Unfortunately, the concepts got blended together, and they have been controversial ever since. If you look closely at corporations today, you will see that some include all three statements, while others only have one, such as a mission statement. To distinguish between them, the definition of each follows.

Purpose: Your reason for being. Some would refer to this as their *calling*. Purpose is focused on you and the gifts you bring to the world.

Vision: How you would like to see the world. Your vision is something that is greater than you and requires other people to make it happen. Vision focuses on the rest of the world and the impact you will have on it by applying your gifts and talents.

Mission: How you live your purpose to bring your vision to reality. Mission is about the process of how you will use your gifts to manifest your vision.

Your values constitute the parameters in which you carry out your purpose, vision, and mission. Values should work hand in hand with your purpose, vision, and mission statements to form the architecture of your career plan. For example, if you value honesty, you will seek out and make deals that have that quality as a foundational element. If you value unity, you will make the effort to ensure your business and your work environment are structured in such a way that everyone is pulling together for the common good. In order to determine your values, you need to answer the question of "why?"

What is your "why?" Let's take the company of Apple® as an example. Upon returning to Apple in 1997, Steve Jobs had a renewed drive to structure the corporation around a well-defined purpose that would define the very essence of why it existed. He made this goal clear during an internal meeting when he unveiled the "Think Different" campaign.[19] At one point in the video of his speech, which I recommend that you watch, Jobs said with conviction, "We've got to let people know who we are and why we exist." According to Jobs, the reason for Apple's existence is that "[they] believe that people with passion can change the world for the better . . . and those people who are crazy enough to believe they can change the world are the ones who actually do."[20] Does that statement tell you anything about the processing speed of Apple devices? No. Does it say anything at all about their primary product of computers? No. The value Jobs saw for the company was far above and beyond its base function. The purpose statement "to reinvent the future" both wrapped up all components into one simple value and released the creativity and innovation of the company's entire staff toward reaching beyond limits or changes.

Take some time now to write your vision, mission, and purpose statements. Don't worry about getting them perfect. The idea is to start and gain more clarity as you progress. Even the biggest corporations hire experts to help craft these statements. Progress, not perfection.

Purpose Statement: _____

Mission Statement: _____

Vision Statement: _____

UNDERSTANDING YOUR SKILLS

Another important aspect in realizing your professional success is having a clear understanding of your skills. Although you may have a sense of what your strengths and weaknesses are, they may well only be general and vague. Having an unclear definition of your skills can lead to burnout—a common occurrence in the business world. By clearly distinguishing between the skills that motivate you and those that lead to burnout, you will be better able to make more productive adjustments to your career path. The idea is to choose those skills you most enjoy and avoid those you dread, then build your life around your motivating skills. This reduces the chance of burnout and increases the ability for you to live according to your highest values.

HARD AND SOFT SKILLS

In order to define and choose your skills correctly, you should first understand the types of skills. There are two basic kinds of skills: hard and soft. Hard skills are those that are objective and tangible, able to be seen, taught, and measured. These are found in such areas as creating and building, analyzing, and problem-solving. Soft skills are more intangible, subjective, and less clearly measured. The best way to clarify the difference is to think of hard skills describing *what* you do, while soft skills are concerned more with *how* you do it. For example, if you are a computer programmer, you utilize a number of skills such as managing projects, applying code language, and testing procedures. However, to complete the job effectively, you must also communicate with your client to understand and plan the project as well as communicate with team members and direct them effectively in order to complete tasks and solve problems. This entails the soft skills of communication, negotiation, leadership, and personal mastery, all of which help you relate to others and carry yourself in a professional and successful manner.

Although both hard and soft skills are extremely important in the success formula, it is the soft skills that will carry more weight as the technological revolution presses forward. One reason is that robots are better geared toward efficiently performing hard skills. Focusing on the development of soft skills is also an area where you can excel above the majority of other humans. Why? Soft skills are underdeveloped because we don't teach and train for them. In one study involving twenty thousand new hires, it was discovered that 46 percent failed at their jobs within the first year. Of those failures, 89 percent involved issues with soft skills or attitude, which included having a bad temperament or a low level of emotional intelligence, being uncoachable, and lacking motivation. Only 11 percent of the overall failures were attributed to a lack of technical hard skills.[21] Clearly, soft skills are incredibly important for success.

This brings us to a division of measurements more commonly used in the technologically transforming world of business, where character as well as smarts defines a successful or unsuccessful person. These two measurements are IQ (intelligence quotient) and EQ (emotional quotient). You are probably familiar with IQ tests that gauge a person's level of mental intelligence. EQ, on the other hand, measures a person's emotional intelligence and has been found to contribute significantly to the success formula. The idea of emotional intelligence was made prominent in the 1990s by Daniel Goleman, who wrote a fantastic book by the same title. Goleman realized that the value of IQ for success had been overestimated. One of the sources for Goleman's argument was a 1921 study by Stanford psychologist Lewis Terman, who observed 1,521 children with high IQ scores. The original expectation was that Terman's kids would soar to great heights due to their high intelligence levels. However, throughout their lifetimes, they did not fare any better than those people with lower IQ scores.[22] This prompted Goleman and others to begin researching the capacity people have for emotional intelligence and what role this plays in their success. Following studies have shown that EQ measurements provide a better analysis of job performance and leadership capacity.

Even though soft skills and EQ are vital for success, it is important to maintain a balance as you strive to develop a skill set that is highly competitive and sought after. The main point is that the business world is changing, which requires the ability to adapt to meet new and developing needs. If you truly want to thrive in the midst of near-constant disruption, you need to set yourself apart from the rest and provide what industries, companies, clients, and customers are seeking.

Take a few moments and make a list of your skills. Rate how good you are at each of them on a scale of 1 to 10. When you're done, next to each skill write an

M if it is a motivating skill or a B if it is a burnout skill. Separate your motivating skills and burnout skills. Make a chart like this:

Motivating Skill 1: _____

Motivating Skill 2: _____

Motivating Skill 3: _____

Motivating Skill 4: _____

Motivating Skill 5: _____

Burnout Skill 1: _____

Burnout Skill 2: _____

Burnout Skill 3: _____

Burnout Skill 4: _____

Burnout Skill 5: _____

You'll want to concentrate on developing your motivating skills. Do more of what you love to do. Some of them you may enjoy, but maybe you're not very good at them yet. That's OK. You can focus on developing those skills. For burnout skills, you'll generally want to do work that involves fewer of these skills. However, some of them may still be necessary. You'll still need to develop a basic proficiency in these skills, but ultimately, the goal is to understand them well enough to find partners who enjoy doing them. Essentially, you should try to outsource your burnout skills.

DO A MARKETPLACE ANALYSIS

It is important to stay abreast of business trends as you go about managing your career—whatever it may be. In an article in *Business Insider* titled "Here's What Banking and Money Will Be Like 30 Years from Now," Geoff Williams gave good insights into the future of banking.[23] While the article did not explicitly address career development, it shone a light on the implications concerning jobs and career growth in the banking industry, if you think about them. While I use this article and the banking industry as an example, the same can be seen in careers in other fields. There is a lot of talk about the changes that businesses experience and even "the future of work" from a corporate perspective, but much less talk about what all of this means from a personal perspective that helps you as an individual.

In the article, Williams used an example of certain banks that are in the process of testing 3-D video banking systems. He also discussed historical changes that the banking sector has gone through, from in-person service to the rise of ATMs and internet banking to our current shift toward mobile banking. While these new services offer customers a better banking experience, people currently in the workforce as well as people newly entering the workforce need to be aware of current trends and how they affect jobs—perhaps even your own job and career. While it is difficult to say if these changes produce or reduce the number of jobs overall, it is clear that the types of jobs afforded by these opportunities are different. Take, for instance, the rise of ATMs. They saved banks money by cutting down on the need for bank tellers, staff hours, and even branches. However, more people were needed to build, deliver, install and maintain the ATMs. While some jobs disappeared, others were created. The result was a shift to different jobs that required people with different skills and interests.

This simple example illustrates the need to remain aware of business trends and how they can affect your career—You Inc. For example, how will the rise of a new technology influence your current job? Will it make your work easier? More demanding? Will you need additional training? Will it reduce or eliminate the need for your position? Will it create a better position for you in your company or at another company? Will it open other opportunities you may not yet be aware of? Chances are, you are already aware of many trends that affect your industry. You may even be making plans within your own company to follow and take advantage of these trends. But how often do you actually take the time to step back, identify current or rising trends, and then use that information and apply it to your career plan?

PREPARING FOR THE FUTURE

To prepare for the jobs of the future or the businesses of the future, the easiest path is to keep up with trends in technology and business. What the first uses of the technologies or business models are doesn't matter—they will spread to other uses and other industries. Maybe you will develop a new technology or a new use for a new technology, end up using a new technology at work or home, or have a new technology destroy the company you work for or take your job. There are too many technologies advancing too quickly for you to be an expert in all of them or even be aware of all of them. That is not an excuse, however, to give up. Keep up as best you can with the time you have available. You might have a billion-dollar idea, see the technology coming that will put your company out of business, or start to learn the technology that will land you your next job. Here is a small taste of some technologies. Seeing even a partial list in one place really demonstrates how many separate innovations are happening at the same time. Many of these are already starting to find their way into business and commercial applications.

Artificial intelligence
Blockchain
Cryptocurrency
Drones
Quantum computers
Robots
Space industries
3-D printing
New materials

What are some of the market trends that will affect your industry or job category?

Market Trend 1: _____
Market Trend 2: _____
Market Trend 3: _____
Market Trend 4: _____
Market Trend 5: _____

Now it's time to bring the information about yourself and the market conditions together to understand how well prepared you are for what you already know is coming. To do this, it is time to conduct a SWOT Analysis, which explores your **strengths** and **weaknesses** and the **opportunities** and **threats** of the marketplace.[24]

DO YOUR OWN SWOT ANALYSIS

The SWOT Analysis has been incredibly popular in the corporate world since it was developed in the 1960s. It is one of the most famous elements of strategic planning. The SWOT Analysis has four elements arranged in a 2 x 2 matrix:

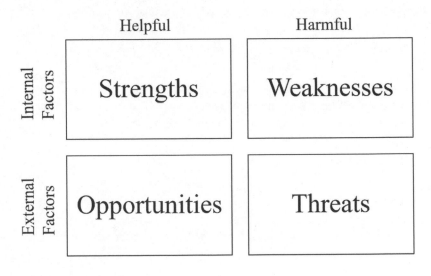

Fig. 8.1.

If you look at the analysis vertically, you see two columns: Helpful and Harmful. If you look horizontally, you see two rows: Internal and External. These create the quadrants of the analysis. Most people don't think or take the time to conduct such an analysis of themselves. Those things that are helpful to you internally are your strengths. Those things that are helpful externally are opportunities. Those things that are harmful internally are weaknesses, and those things that are harmful externally are threats.

Strengths

Regardless of what is happening in the outside world—e.g., AI, robots, the economy—you must determine what your strengths are. Remember, your strengths are more than just your hard skills that can be learned. They are who you are, which brings up important questions. How can your strengths be used in the world while it's rapidly evolving? If a robot is likely to take your job, where else can your strengths be used?

Weaknesses

It is time now to consider your weaknesses. It is important to understand your weaknesses and whether you need to focus on addressing them to achieve a high enough level of competence to accomplish your vision, or whether you should hire or partner with someone whose skills complement yours. Or perhaps you should ignore your weaknesses completely and choose another direction where they are not relevant to your success. Now, write down your own analysis of the weaknesses that exist in the marketplace concerning your field of expertise.

Opportunities

Take a look around you and identify the opportunities that exist that you can take advantage of. Think more broadly than just the next job you want. As illustrated by the banking example, what opportunities would exist for a bank teller in a world that is turning to mobile banking? There are always opportunities. You simply need to look for them. It is during times of great change that the next generation of success arises. Even during the Great Depression, there were incredible opportunities for some people to improve their positions if they applied themselves. For example, during that era, most companies cut their budgets in order to protect themselves. As a result, many of them reduced their advertising budgets down to nil. Proctor and Gamble, on the other hand, took advantage of the fact that its competitors downsized their advertising efforts, and expanded its marketing. Instead of going with the flow of business practices of the day, the company made use of the vacuum that those created. As a result of this bold approach, not only was Proctor and Gamble not affected by the depression, but it came out of the crisis with incredible brand recognition.

Are there situations you can take advantage of? Are there partnerships that could be formed that would help you realize your goal? Are there openings in the marketplace that present advantageous stepping-stones? Now, write down your own analysis of the opportunities that exist in the marketplace that can be taken advantage of according to your unique promise of value.

Threats

Last, it is important to consider what possible threats there are to your vision. Just like opportunities, threats to your efforts always exist. For example, your particular career may be threatened by an occurring or possible upcoming shift in the industry. Perhaps there is some environmental challenge that will make your vision difficult to establish. Or perhaps the success of your venture will mean that someone else's business will suddenly become obsolete, causing that person to work against you to prevent your success.

Is the marketplace surrounding your endeavor particularly volatile? Is there some new technology that could make your project obsolete? Are there existing industries that will be threatened by your innovation that might want to sabotage your work? Now, write down your own analysis of the threats that exist in the marketplace to you.

The point is this: You are the owner of your career. No one else. You are the one responsible for the results you achieve. As the CEO of You Inc., you have to ask yourself this question: what are you doing to prepare your business to move forward and thrive?

Now, we will apply the SWOT Analysis to You Inc. To do this, start by reflecting on your vision statement. Remember that your vision statement is a compelling view of what you want the world to look like through living your vision, mission, and purpose. Consider the landscape around you today, and ask yourself what will help you and what will hinder you as you proceed with your mission.

My strengths are: _____

My weaknesses are: _____

My opportunities are: _____

My threats are: _____

The information you've gathered about yourself and the marketplace is critical to remaining employed and having a financially secure future. Before developing actionable steps you can use (see chapter 10), we'll use the next chapter to look at other ways you can maintain your financial security from the viewpoint of your savings.

CHAPTER 9

REMAINING FINANCIALLY SOUND IN DISRUPTIVE TIMES

n chapters 1–3, we developed an understanding of what an industrial revolution looks like, the never-ending quest for automation and how the workforce is being redefined. In chapters 4–7, we looked at the nature of the disruption, exponential change, and the forces shaping the future economy and labor market. In chapter 8, we looked at how to maintain an income by changing your mind-set, more deeply understanding yourself so you can be more agile, and getting prepared to ride the wave of change. Being employed, whether through a full-time job, freelancing, or starting your own business is only part of maintaining financial security—having an income. But what would you do if you knew, without any doubt, that three years from today you would lose your job? What actions would you start taking today?

Having income is only part of what you need to be thinking about. What about your savings? Insurance? Healthcare? Home? Well, it's time to put on your chief financial officer hat. And you might as well bring along your chief risk officer, chief investment officer, chief technology officer, chief human resources officer, directors of accounting and tax, accounts payable, accounts receivable, treasurer, and your bookkeeper.

From an accounting perspective, there are two types of assets—tangible and intangible. Tangible assets are assets you can hold or possess (to touch). Your tangible assets may include bank and brokerage accounts, real estate, cryptocurrencies, precious metals, cash, and other similar items. Your intangible assets are assets that you can't hold and include your knowledge, intuition, personal brand, and so on.

As you move deeper into the disruption of the fourth industrial revolution, you also need to consider what will happen to your assets and how to manage them through disruption. In fact, you may need to rely on your assets to carry you through periods of disruption, like a job loss or a health problem. Often, we manage our assets on autopilot, if at all, just following some rule or advice we've heard. Invest for the long term and forget about it. Save six months of salary for an emergency fund. Save $3 million in your retirement accounts. But if you think about it, if jobs, companies, and industries are being disrupted, there's a good chance that every aspect of your financial life is being disrupted too—your bank, your insurance, the way you pay your bills, /and your savings and investments. Managing your tangible and intangible assets amid new technologies, disruption, and exponential change will require being aware of the options available to you as well as knowing how and when to act.

If that weren't enough, we need to consider how life will change as a result of these new technologies. For example, advances in medicine and genomics may lead to extending life by decades. Our understanding of retirement will fundamentally need to shift, even for those who are already retired and who may benefit from these advances. Will your needs for life insurance change? Will you be able to find a job at age seventy if that's new middle age? Can you save enough for a seventy-year retirement? What will happen to social safety net programs for the elderly?

These are questions no previous generations have had to ask. While these miracles are not all here today, they are also not so fantastical that they are only the stuff of science fiction. You should be thinking about what you'll do if you live longer. Much longer.

The good news is that dramatic changes haven't yet arrived. There is time. But you also don't want to fall victim to "boiled frog syndrome," where it's said if you put a frog in boiling water it'll jump out, but if you raise the temperature slowly it will get used to the gradual change and end up boiled. It takes time to understand the changing dynamics, what they mean to you, and what you need to learn to thrive in a new machine age. There is no better time than right now to begin.

QUESTION WHAT YOU THINK YOU KNOW

Traditional financial advice suggests that you should have an emergency fund of savings that you could use to pay your bills for a certain amount of time if you lose your job. How much savings depends on who you ask, but generally it is in the range of three to twelve months. Why that range? Traditionally, that's how long it takes to find a job if you lose yours during a typical economic recession in the post–World War II era. When a downturn occurs, a manufacturer might slow production or shut down a line until demand picks up and you are hired back or perhaps find other employment.

But what happens when an entire industry is shaken? Multiple industries at the same time? What happens when technology is changing rapidly at the same time so that when the economy picks up again, there is no need for you anymore? The Great Recession offers some insight. The chart below (fig. 9.1) shows the average number of months that a person who loses his job can expect to be unemployed. Note that this is the average, so some people find work again more quickly and others take much longer. Note also that the *average* time of unemployment almost doubled! And, almost a decade after the peak, the average time of unemployment is still higher than at the peak of any recession since the end of World War II. When the wave of automation and other disruptions starts to happen, those who lose their jobs will need to be ready for a long period of unemployment, very likely years.

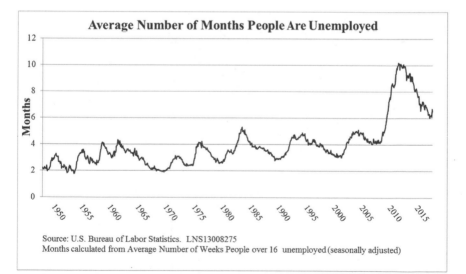

Fig. 9.1.

The next chart (fig. 9.2) shows how many people are considered long-term unemployed—unemployed for twenty-seven weeks or more. From the perspective of economic policy, an economist would normally consider this number in relationship to the current size of the workforce. Naturally, the population and number of people in the workforce tend to increase with time, so there should be a natural drift upward from this perspective. For the economy as a whole, this number of long-term unemployed people may not be a problem, particularly as robots and algorithms continue to keep companies humming along. However, we are not talking about the economic policy of nations but rather the Economy of You. And you need to be aware of the environment you are likely to be in— with millions of others.

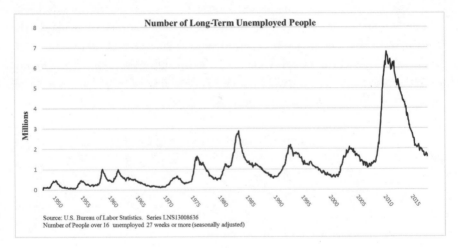

Fig. 9.2.

Your perspective and response will depend heavily on your own circumstances, many of which are age related. What is your career stage? How close to retirement are you? How much savings and debt do you have? What risks should you be reasonably concerned about now and in the future? These answers to these questions will be loosely tied to your age. For ease of discussion, we'll look at the issues in terms of generations—baby boomers, Generation X, and millennials.

Baby Boomers: As a group, you are either retired or approaching retirement age in the coming decade or so. Some of you may make it to the end of your careers without experiencing the direct impact of automation. Others will have automation bring an early end to your career or find it difficult, if not impossible, to find a new job—or even learn the new technologies to get a job. Traditional issues of age discrimination will be compounded by much of the value of your work experience being discounted. Younger workers who have grown up with technology, have business acumen, command a lower salary, and have fewer life issues holding them down will increasingly win out in the job market. Of course, some of you will manage to navigate the change fine, but like a game of musical chairs, even if you are good, there won't be a place for everyone. Boomers with savings will have opportunities to invest in future technologies and reap the rewards. Others may be able to start their own businesses. As a boomer, you need to be thinking about what you will do if you lose your job and cannot find a similar-paying job for years, if ever.

Generation Xers: Your job and your career will be affected by automation and disruption from technology and business models. The good news is that you've grown up with technology and some disruption. The bad news is that the technology of tomorrow will be different, and to fully take advantage of it, you'll need to learn to think differently. You are either in the prime or approaching the prime earning years of your career. Your future will not look like what your parents' future did. Of course, you always knew that anyhow.

What the impact will look like will depend on what your career is. Executives are unlikely to have their roles automated but will need to make decisions about using new technologies to stay in business and be aware of competitors coming from other industries or seemingly out of nowhere. Employees will need to be aware of and learn new technologies to both use them and be aware of threats to their jobs and the company's viability through change. Executives and employees alike need to be aware of what options there are when a job loss comes their way. It's critical to understand that when it comes to new technologies, chances are you have less knowledge and are more expensive than younger job seekers. How will you compete for jobs and business?

Now is the time to get your financial house in order. Reducing debt, increasing savings, and investing in developing new knowledge and skills will be critical. More than any other generation, Gen Xers are generally in a precarious place, able to capitalize or be swamped by the wave of change the fourth industrial revolution will bring. You need to be paying attention to developments and strategizing how to respond and stay relevant. Learning to partner with younger Gen Xers and millennials to leverage your business savvy could lead you on a successful path. You need to have a plan B.

Millennials: The oldest of you are starting to make your impact on the world, while the youngest are still in school. The good news is that your generation is driving a lot of the change, and you have learned and are learning the skills of the future. This does not mean navigating the fourth industrial revolution will be easy, especially if you are saddled with educational debt. Being early in your career, you have the advantage of less to lose than the older generations when technology shifts. You are in a position to take more risks and to bounce back from failures all the wiser. You too will need to keep an eye on technological developments and continue to play with and learn new technologies. Key for your long-term success is developing leadership, management, and business skills. Look to partner with older generations to leverage their knowledge and skills to help bring your ideas to life.

BUILDING YOUR TANGIBLE ASSETS

With so much change going on, it's difficult to know how to prepare your savings and investments and adjust your strategy. This is primarily because most of us have been told a common story about savings and investments that was sound, average advice for the mid-twentieth century. You can still follow that advice into the fourth industrial revolution, with some modification and special attention to the companies that will benefit from and fall to disruption.

You need to both challenge what you know and learn about other possibilities for your investments. In his book *Killing Sacred Cows: Overcoming the Financial Myths That Are Destroying Your Prosperity*, Garrett Gunderson challenges conventional financial planning wisdom and shows that there are different, more secure ways of managing your personal finances with existing technologies and strategies.[1] Is your 401(k) plan a trap? Is there a better alternative? What's the difference between price and value? What was good in the past is not necessarily a good strategy today or tomorrow.

But what are the alternatives? They aren't taught in schools. Most parents don't even know what options are available and therefore can't teach their children. Generational wealth is more than just the passing of money from one generation to the next; for some families, it's teaching and a way of life. In the following section, we take a look at some of these alternatives. It is important to keep each of the possibilities in mind from two perspectives—as an investor looking for a return and as an entrepreneur or a business owner seeking investment money to grow. It's important to understand enough about each so you can decide which are the best possibilities for you so that you can learn more about them. Putting money into an asset or process you don't understand is gambling, not investing. Take the time to learn. Spend some money on books, training courses, and mentors as needed. Your investment in knowledge will pay off in either making better investments or avoiding costly ones.

THE DEMOCRATIZATION OF INVESTING CHOICES

One of the power benefits of new technologies, business models, and regulations has been the democratization of investment options. No longer are you limited to stocks, savings bonds, certificates of deposit, money market accounts, and savings accounts. If you are starting a business, no longer are you limited to money from family or billionaire venture capitalists. Here is a sampling of some of the options you have available to you today.

Crowdfunding: This is a means of raising money for your business or project from the general public, rather than from venture capitalists, angel investors, banks, or other sources of large amounts of money. Crowdfunding was legalized by the JOBS Act of 2012, though it existed in some form prior to that.[2] Entrepreneurs and small businesses looking for money to build out their projects simply post a description of what they are doing on a crowdfunding site and tell donors what they may be getting in return for their contribution. The three types of crowdfunding are reward-based, donation-based, and equity-based. Reward-based crowdfunding usually offers the product or bundles of the product as a reward for contributing. Donation-based crowdfunding is seeking financial donations with no compensation in return. Donation-based crowdfunding is often aimed at charitable causes such as natural disaster relief or raising funds for a worthy purpose. Equity-based crowdfunding gives part ownership of the business to contributors.[3]

Regulation A (Reg A): This SEC regulation allows small businesses to raise money without having to adhere to the same level of disclosure requirements and filings as larger companies when issuing securities. For those seeking funds, there are two tiers of investments. Tier 1 allows for receiving up to $20 million in a twelve-month period. Tier 2 allows for up to $50 million to be raised in a twelve-month period. The pool of potential investors is also opened up, and nonaccredited investors may invest a certain portion of their income into a Regulation A offering.[4] Whether you are looking to raise money or invest money, it is important to understand the regulations and seek appropriate counsel to ensure you don't run afoul of securities law. Don't wait to start learning until you are ready; start now so you are ready if an opportunity or need arises.

Initial Coin Offering (ICO)[5]: With the recent rise of blockchain or distributed ledger technology, many companies are offering digital "coins" as a means to raise funds. ICOs are in their infancy and have yet to be regulated to ensure there are financial safeguards for investors. A number of countries have banned ICOs altogether, while others are cautiously watching. Companies that offer ICOs as a means to raise funds will typically have a white paper explaining their project and a road map of targets they plan to hit. Once funds are raised, the coins are distributed to contributors on one of the cryptocurrency exchanges, or in a wallet you can store on your computer or in another safe location. Like any start-up, the company offering the ICO may do very well, or it may crash, making the coin worthless. Now is a great time to start educating yourself about ICOs, how they work, and how you might invest in them or raise funds through an ICO. Be careful, as you can lose your money or run afoul of securities rules and tax law.

Angel Investing[6]: Another way to raise money for your business or invest your money for a return is angel investing. Angel investors are interested in helping small businesses launch.[7] While they do expect a return on their investment, they are often investing with the hope of helping the business succeed. Angel investors use their own money and generally need to meet the requirements of an accredited investor, though there are times when nonaccredited investors also participate. More people are eligible, or one day will be eligible, to be angel investors than they realize.[8]

Peer-to-Peer Lending: Another way of investing your money or receiving money for your project is through peer-to-peer lending. Rather than going to a bank for a loan, it's possible to borrow money from individuals who have money and are looking for a return on investment that is greater than their low rates on saving accounts, certificates of deposits, and even stock market investments. There are platforms that can help facilitate matches between potential lenders and borrowers. Generally, these loans tend to be small—microloans. Of course, there is a risk the business will fail and the loan will never be paid back.

Self-Directed IRAs: These have more options available for you to invest in beyond the typical mutual funds, stocks, and bonds offered by most financial services firms. With a self-directed IRA, you are able to invest in virtually anything—rental properties, precious metals, gems, cryptocurrencies, and even as an angel investor in early-stage start-ups. It is a plus if you can find a custodian to manage your account holdings for you. You can direct where and how funds are spent while the custodian handles the transactions, record keeping, and holding of any assets and documents. Self-directed IRAs are more expensive to maintain and so have higher fees, higher risk, and potentially higher returns.

INVEST SMARTLY

A strong investment portfolio has always been a foundational pillar of security in changing times. In the past, the investment process was simpler, requiring that you simply identify the rising giants and invest in their efforts. However, as the world became more global in nature and technology opened the floodgates to more investment options—both lucrative and destructive—wise investment practices moved from the singular to the diverse. In order to make up for more and greater potential losses, a wider swath of investments was needed in one's portfolio. Adjustments had to be made to the old way of investing to incorporate a growing number of risks occurring on a broader business playing field.

The necessary changes to investment practices are a good example of how to handle uncertainty in a rapidly changing world. The entire process was not thrown out like the proverbial baby in the bathwater but, rather, was restructured to meet new challenges. Investment remains a pivotal cog in the wheel of change, but investment procedures must be tweaked yet again in order to better meet approaching disruptions. Since technology lies at the base of coming change, it is the companies and organizations that most effectively develop and/or utilize it that provide the best ROI and, therefore, the best chance of remaining economically sound as technology rises and other trusted aspects come crashing down.

Once you identify factors that open investment opportunities and drive profits, you can begin to search for those companies and organizations that offer the best ROI during times of transformation. We've already identified advanced technology and a faltering economy drowning in debt as the main drivers of change, which greatly narrows down investment search options. According to these factors, there are certain traits that potential investment sources should possess or be actively pursuing. Critical criteria to be considered include:

Problem-Solving Efforts: Is the company or organization in which you are interested taking steps to actively address problems using viable methods? As we have discussed, many corporate heads, boards, and committees seek to apply outdated and ineffective methods to rapidly changing problems, which will ultimately end in their demise. However, those places that are approaching the changing times with revolutionary techniques designed to work with, not against, technological advancements have the best chance of thriving during the transformation. The process should involve the active participation of customers, since it gauges how the public is responding to new techniques.

Forward-Looking Plan of Execution and Qualified Management Team: The company or organization should have a well-defined plan of execution that moves it from current performance standards to sustained performance output as future trends morph into reality. Although such a strategy is key, it is only as good as the management team that implements it. Therefore, you should also ensure that a qualified management team is in place that includes not only people who are experienced in modern trends and practices, but those who possess the vision to see into the world that lies ahead.

Diverse Capability Expansion: Is the company or organization investing in the expansion of diversified capabilities? Just as diversity played a major role in corporate investment when it expanded in earnest globally, it plays an even greater role in a successful transition from a traditional to technological business model. Companies that seek to expand their capabilities beyond what is necessary and invest more broadly in information technology and robotic systems will gain a market edge.

Implementation of Technological Systems: If companies and organizations are expanding their capabilities, do they have the capacity, personnel strength, and means to successfully implement technological systems and projects? It is one thing to desire and plan to convert to or deploy new technology, but it is quite another to be in a position to actually implement those desires and plans.

Clear Value-to-Risk Ratio: A company or organization that offers good investment potential should possess a clear value-to-risk ratio. Risk is a natural and present part of the investment process, but you want to ensure that the value of your investment is equal to or greater than any risk involved. Establishments that successfully forge ahead are usually those that have a clear and realistic understanding of how their value positively outweighs risks to their investors.

Growth Potential: The company or organization you invest in should not only provide a clear value-to-risk ratio but also demonstrate that it has what it takes to achieve high growth—especially through the challenges of rapidly changing times. It must be able to show that it is capable of recognizing and engaging market opportunities as well as demonstrate the ability to understand and maneuver the landscapes of both customer bases and competitors within its markets.

Sustainable Competitive Advantage: Ideal investment targets will already be ahead of the curve in both potential and actuality. This includes having aggressive upgrades in process and drawing the attention of new, well-funded investors who share a futuristic vision. Such companies or organizations will have relatively few competitors that are as big as or bigger than they are. This type of business climate provides the investment target with the greatest potential for obtaining a competitive advantage that maintains sustainability in a rapidly changing, volatile global market.

Return on Investment (ROI)—Of course, the bottom line of dropping money into any company or organization is to achieve a good, even great, return on investment. This is particularly crucial during the kind of turbulent and uncertain times into which it appears we are free falling. It's one thing to lose money during stable times when losses aren't so upsetting, but when conditions are volatile, the loss of investment capital can mean quick defeat that leads to extended suffering. Ensuring that the previous points are present will help to also ensure that you receive a good ROI as the technological, economic, business, and social landscapes change.

ROTH 401(K) AND ROTH IRA

Roth retirement plans offer some advantages that are important to consider. Roth plans are different from traditional 401(k) and IRA plans in that your contributions are made after taxes. Since you have paid taxes on the amount you contribute, there are two important implications of using these plans in the context of savings during disruptive times.

1. Contributions can be taken out without penalty—per rules of the plan.
2. Withdrawals on growth are tax-free when you are eligible to withdraw.

The first point on taking out contributions without tax or penalty can be beneficial if you lose your job and need some money to pay the bills. You've already paid the taxes, so you can withdraw the money. The important issue to understand with this is that if the money is in a Roth 401(k), your company's plan rules may not allow you to withdraw the money while you are employed. In such a case, if you are using the money because you lost your job, simply roll over the Roth 401(k) to a Roth IRA and proceed. This will take some time. It may take about four weeks after termination to be allowed to do the rollover, and a rollover may take a couple of weeks as well. If you keep the money in a plan, you may be able to make a hardship withdrawal too, but it's best to do this from a Roth rather than traditional plan as well. Withdrawing from a traditional plan always involves paying taxes on the withdrawal. As a silver lining, maybe your tax rate is lower after losing a job. The point here is to have options available to you to choose from if or when the time comes. Of course, you should consult with your tax or personal financial advisor to understand the details and weigh your options.

The Roth plan's tax-free growth is also important if you are investing in high-growth businesses or other assets. If you contribute $10,000 and it grows to $100,000, you paid taxes on the $10,000 and get to withdraw the $100,000 tax-free. With a traditional plan, you don't pay taxes on the $10,000, but you do pay taxes on the $100,000 as you withdraw it. If you're investing in riskier, potentially high-return assets in your retirement savings, this would be an important consideration. Of course, you should be consulting with your tax or personal financial advisor to clearly understand all of the risks and benefits.

INVESTING IN YOUR INTANGIBLE ASSETS

In the age of artificial intelligence, robotics, and automation, it's your intangible assets that separate you from the machines. Your intangible assets are your skills, your knowledge, your reputation or personal brand, and so on. Some of these assets are depreciating, meaning they become less and less valuable, maybe even worthless. For example, being skilled at typewriter repair has little value in today's world of computers and smartphones, even though there was a time it was a lucrative skill to have. Be sure to be investing both time and money in developing and increasing the value of your intangible assets. People make the mistake of believing they don't have to invest or that someone else, like their employer, should invest in their development. Your financial well-being depends on you investing in yourself.

INCREASE LEARNING

Education is another field in which old ways and methods are being clung to while major changes make much of what and why we learn redundant and inadequate. Emphasis is often placed on the act of getting a degree—any degree—without much guidance as to which degrees are best suited for a rapidly changing world. Most elementary schools, high schools, colleges, and universities still utilize the model of seating students in orderly rows so that they can learn to obey commands of authority, perform well on specially constructed tests, and think and act mechanically. This form of education served society well when machinelike performance by humans was required to fill large numbers of monotonous manufacturing and office jobs.

However, as we have seen, the world around us is changing at breakneck speed, and the way in which education is implemented must be radically altered as well in order to keep up. Jobs requiring repetitious actions are being taken over by robots and other forms of high technology, and many of the degrees that students pursue with the goal in mind of simply "getting a degree—any degree" are becoming obsolete and will soon fall victim to robotics and the revolutionary changes they are inspiring. It is even worse to skip a college education in the hope that sustaining employment will be found without it. This also used to function fairly well for a high percentage of students who ended their educational pursuits after high school, but it was during a time when society had masses of manufacturing and service jobs that didn't require a degree. Today, those kinds of jobs are rapidly disappearing or being filled by robots that are cheaper, more efficient, and more productive.

In order to remain economically sound through this type of major transformation, it is critical to alter educational goals. If you are seeking to obtain a degree, consider fields in STEM (science, technology, engineering, and mathematics), which are all involved in the creation and development of new technology. If you aren't a "university type" or can't afford that level of expense, consider trade schools that provide skills relevant to the changing business landscape.

Bottom line—increasing education and knowledge about cutting-edge developments and technologies will give you an advantage over others who fail to remain current.

IMPROVE SKILLS

You may have already obtained a degree or other training in a field relevant to computers and robotics and be happily involved in a fulfilling career. However, it is prudent to not stop there but to continue pursuing learning in that field as well as expand your knowledge and skills in related areas. There are a couple of important reasons for doing so. First, you maintain a knowledge and skill level above the competition, which will make you stand out for both current and future jobs and projects. Second, what you see today is bound to change tomorrow as exponential growth continues to transform the business and social landscape.

As the web of technology continues to grow and become increasingly interconnected, it is crucial that you also obtain knowledge and skills that are related to or even parallel to your particular field. For example, a job that requires maintenance on a certain type of machine often also requires that you have at least a surface knowledge of the computer system that operates it. If you currently hold a position in hotel management, seek courses in a field such as IT. Parallel careers are growing in popularity, particularly with millennials, who are being faced with an increasingly fast-paced morphing of job markets.

Just as with financial investment, you should diversify your personal education and training portfolio. That way, if major changes arrive and affect your present job, you can sidestep to another career for which you have been preparing. Parallel career training and skill honing are also smart because the business world is more volatile than it used to be. More companies are being forced or are choosing to cut human employees in order to maintain competitiveness— and many of them are turning to robotic systems as replacements. If your management or administration job is threatened, you could move to a position in IT, computer analysis, or some other position that is in more demand. Oftentimes, new jobs in these areas pay more—or at least have the potential to pay more— than those positions lost.

Bottom line—the more knowledge and skills you have in areas that are emerging or are set to unfold, the more able you'll be to navigate changes successfully.

BE A RESOURCE MAGNET

Resources are a vital aspect of any endeavor. Whether you are part of an army marching off to war, an explorer heading into the depths of unknown territory, or someone with a great idea wanting to start a business, a good and steady stream of resources is necessary to fulfill the mission. If the required resources cannot be gathered at the beginning, or they run out somewhere during the process, failure is imminent. Therefore, the ability to acquire resources, both at the start-up stage and throughout the operation, is a necessary skill if you want to succeed. Every venture requires some degree of capital to get it off the ground, and the greater your vision and expectations, the more venture capital you will need. Once your project is up and running, finances are necessary to meet current operational demands and market your product or service, among other needs. If you desire to grow your business, you will need extra capital to acquire or expand building space, add or update equipment, or hire more people, for example.

Although finances are the lion's share of what is required to successfully launch, maintain, and grow an idea, there are other resources that are just as important. You also need to know the right people to either support you or point you to those who can. It is a fact that you cannot achieve success on your own. Others are needed to keep the wheels turning on multiple levels and open doors you are not yet aware of. "Right people" also applies to those you decide to employ. If you hire people who do not fit properly into your vision, it could end up costing you a great deal of time and money as well as threatening the very success of your objective. You will also require a good and constant supply of those resources necessary to feed your endeavor. If you produce a beverage, for example, you will need a steady supply of such things as water, sugar, additives, bottles, lids, and labels. If your supply line breaks down at any point, your business is threatened.

Nearly every aspect of life today—e.g., business markets, social interaction, and product and idea flow—is continually expanding on a global scale, which has both positive and negative effects. Positively, global communication and interaction open a huge number of opportunities that only a few decades ago did not exist for the average person. Because of the limited ability to travel, communicate, and gain an audience with the right people, the golden deals were reserved for the few who the ability to secure precious resources. Basically, if you didn't know the right people or weren't wealthy, the chances of starting and progressing a business idea beyond the local level were slim at best. However, global access has eliminated a mountain of obstacles, giving average people access to both opportunities and resources and allowing them to start up and expand their ideas and projects like never before.

Worldwide expansion has created some difficult challenges for those 'preneurs seeking to make their mark in the world. Along with the flood of opportunities presented via the opening of global access comes a massive degree of competition. You may find more ways to obtain resources, but so have billions of others. Markets are larger, but there are significantly more people offering the same or similar products or services as you. You have access to more investors, but their attention is being sought by a long list of others seeking their resources.

In order to succeed in an increasingly global world, you must be able to acquire the resources needed to overcome obstacles—and the obstacles are only becoming more challenging as the world moves into the Age of Robots. As robots replace more workers, those displaced people will be seeking other ways to make a living, creating an even greater atmosphere of competition for the resources required to make their ways work. Sheer demand for resources causes them to become scarcer and increase in price when they are found.

You must become a master in the art of attracting the people who have the resources you need if you are to succeed in this time of exponential global growth, and that requires having and effectively applying certain skills. Remember, the field of competitors that is growing quickly is vying for the same resources you are seeking, so suppliers—whether of money, talent, or material items—respond to those who meet certain criteria and offer the best return on their investments.

Some of the qualities that will attract resource suppliers to you over others include:

Having Unique Ideas: There are multiple millions of people who have ideas they want to promote. Shrewd investors, workers, and suppliers scan the sea of people pleading for support in search of unique ideas— or existing ideas presented in unique ways—to put their weight behind.

Developing Value: Resource suppliers also look for ideas and projects that demonstrate the potential for increasing value. If you develop value in your project and can show a real potential for growth, investors and suppliers will be more compelled to back you over others.

Demonstrating 'Preneur Characteristics: When you launch a project, you are actually selling yourself as much as your idea, and those doling out the resources look for certain qualities ingrained within you as they listen to your sales pitch. In order to more effectively attract the resources that you require, you should demonstrate to investors and suppliers that you have the key ingredients of success, which include passion, knowledge, commitment, flexibility, tenacity, and people skills. The better you present a well-rounded 'preneur character, the more likely people will fall in behind you and your cause.

Gathering a Killer Team: Experienced investors and suppliers understand well that no matter how impressive an idea is, it will not succeed without the right team of people involved in making it happen. It is critical to select people who catch your vision and are just as passionate as you for its success. Those people you are turning to for resources know that such a well-assembled team will do what it takes to succeed and will not collapse under the pressures that arise along the way.

HARNESS TECHNOLOGY
FOR COMPETITIVE ADVANTAGE

It is clear that we are at a stage where advanced technology is set to appear in virtually every crevice of society, and there isn't much we can do to stop it. Artificial intelligence systems are beginning to dominate logical reasoning processes, and they are learning to master pattern-recognition processes. AI systems are already beginning to excel over humans in certain areas such as performing mathematical equations, providing solutions to complex problems, performing delicate surgeries, and diagnosing various diseases. Although robots are overtaking humans in a growing number of areas such as task accuracy, task performance, and intelligence, they still lack the ability to make decisions based on morality and ethics. For example, a self-driving AI car cannot stop to assist an elderly lady across the street, and an AI robot providing medical advice cannot offer comfort or emotional feedback to a suffering patient.

Instead of fearing its arrival and complaining about all the possible negatives it brings with it, we would be prudent to harness the power and potential of this new age of artificial intelligence, robotics, and other forms of advanced technology and use these elements to our benefit to create an atmosphere of creative advantage. As we have discussed, the expanding global market is magnifying competition, and we need to access the tools available to us to stand out from the crowd. A juggling act is required to pull this off during times of rapid change and unknown outcomes. Both human and machine potential together need to be identified and maximized to achieve an optimized outcome.

The practice of gaining competitive advantage has been around for decades. The difference is that today's technology, such as AI, is having the same effect as it is on nearly everything else—it is advancing at a rapid pace that makes planning based on linear thinking an ever-growing impossibility. Traditional methods of attracting and keeping customers, distributing products and services, and managing assets are no longer sufficient in themselves to stay afloat in a rising sea of those using advanced technological systems. New technology gives companies an edge, but even that is short-lived because a growing number of organizations are deploying similar technology—or better, as today's technology becomes quickly outdated thanks to tomorrow's developments.

PERSONAL BRAND

Your most valuable asset is you. You are, after all, the source of all income and savings you have and the source of all of it into the future. As we saw in the previous chapter, your personal brand is not just a positioning statement but a deep understanding of who you are and what you stand for in an authentic way. To increase the value of your personal brand, I recommend following Reach Personal Branding®'s three-step approach, described by their 1-2-3 Success!™ process.[9] The three-step process is summed up as "extract, express, exude," the details of which you can read on their website. As your brand is your most valuable asset, you need to keep it growing and developing. This includes working on yourself to better understand the evolving you, reassessing your interests and priorities, communicating to the world your brand and your value, and ensuring that your life is aligned to be authentically consistent. The value that others see you able to create, whether it's monetary or a positive impact on the lives of others, determines whether they are supporters or detractors, customers or not, advocates for you or not. A strong personal brand opens doors for you and enables you to earn a premium over services with a less valuable brand.

SOCIAL MEDIA

Having a presence on social media platforms, one that is consistently aligned with your personal brand, is increasingly important. Platforms like LinkedIn®, Facebook®, beBee®, Twitter®, Instagram®, GitHub®, and a host of others offer many benefits, and those who are reluctant to participate will increasingly find themselves left behind, even viewed as out of touch.

Regardless of what your role is, as an employee, entrepreneur, or board member, social media is the first place anyone from potential clients to employers will go to understand who you are, whether you are legitimately qualified, and what your character and personality are like. This is one reason why it is important to present yourself consistently in brand alignment across social platforms and to be sure people understand who you are. It is a mistake to believe that limiting what you share with certain audiences will prevent people from getting the wrong impression. The reason the people closest to us are often least likely to help us with our business or careers is because they are not professionals in your areas of expertise and have a different non-professional view of you and your life. While you may limit showing the picture of you dancing with a lampshade on your head to your friends, that's the impression they have of you, and they likely won't be recommending you for a job or contract either.

Social media also offers you the opportunity to showcase your talents and abilities. Some sites will allow you to interact with live video, and others to post articles, share presentations, and more. Sharing in these ways helps you both to demonstrate your skills and talents and to be found by people looking for someone who possesses those skills and talents. For those in leadership and thought leadership roles, as well as those looking to move into those roles, social media provides a valuable platform for showcasing yourself and your personal brand. It also helps you to be found when people are searching.

In this age of automation and robots, having a strong social media presence will become increasingly important to your long-term success. Because it takes time to develop a social presence, you need to start right away and consistently be adding value on the sites you choose. Find valuable content that is aligned with your personal brand and objectives, add some of your insights, and share. Show both your personal and business sides. Deeper connections are forged through shared personal interests and curiosity about new ones. Show your value and be human to make personal connections. Don't just add connections; take it to the next level and develop relationships with the people you connect with. Respond to each connection, looking to learn more about them and share more about yourself.

NETWORK

It is said that your network is your net worth. A carefully developed network can be of tremendous help to you professionally and personally when you develop genuine personal relationships. Over time, people will come and go in your network and you in theirs. That is perfectly natural and to be expected. The key to successful networking is to be of value to others rather than seeking how others can add value to you. To paraphrase President John. F. Kennedy: "Ask not what your network can do for you, but ask what you can do for your network." By providing value to others, you will be seen as the go-to person for however you help people. Whether you solve IT problems, answer questions about grammar, or clarify what future technology means to everyone, the value you provide can lead to opportunities—from a new job to starting a business to finding a difficult-to-find spare part for the car you are restoring.

Start to think of everyone you know as part of your network. You can even break your network down into categories of what their expertise and value are (their personal brand). Who are the experts in analytics, social media, artificial intelligence, cryptocurrencies, doll collecting, or underwater rugby? This way you know who to turn to when you hear of opportunities as well as when you need some help.

Most people fail to sufficiently develop their network and keep in touch. Then when it comes time that they need some help, they find themselves not knowing who to turn to or feeling embarrassed to ask for help when they haven't offered any value themselves. The time to get to know the people in your network is now and every day, long before a robot takes your job or you start out on your own. Networking expert and strategist Andy Lopata summed up the first step to successful networking: "Connecting is not enough."[10] Depending on how you meet people, you will want to crossconnect with them online and in person. Email, messaging apps, texts, phone calls and in-person meetings are great ways to get started and develop your relationships. Consultant Sarah Elkins is an example of someone who creatively took her networking to the next level, developing a conference, No Longer Virtual™, for people to take their virtual connections offline and meet in real life while also getting the benefits of business-oriented trainings. There are many creative ways you can create your network and deepen your relationships.

REMAINING FINANCIALLY SOUND IN DISRUPTIVE TIMES

In an age of increased automation, it becomes increasingly important to understand and develop both your financial and your personal assets. You need to develop them both for defense, in case you lose your job, and for offense— to grow, develop, and successfully compete and work with new technologies. Robots, algorithms, and automation can be used to enhance your assets, to help you be better at being *you*, and to provide you with financial stability. To do that, you need to understand them and their abilities and get creative on how to use them for your benefit.

CHAPTER 10

GETTING OVER IT
AND GETTING STARTED

The amount of change occurring in technology and business can seem overwhelming. The robots and algorithms are making inroads; change is happening exponentially; and jobs and even entire industries are being created, destroyed, or transformed into something new. Keeping up with it all can be exhausting and overwhelming. What do we do?

As we've seen, you will not be able to outcompete a robot at a task it was designed for. Robots will be faster, cheaper, and less likely to make a mistake by the time they move from the lab to the workplace. The answer to succeeding lies not in competition but in cooperation. Finding ways to work with and use robots and other new technologies do require putting in the effort to learn them. Technophobia and taking pride in not knowing the latest technologies is a recipe for the unemployment line and possibly losing your investments. The robot in the next cubicle is here to stay.

In an increasingly winner-takes-all economy, the most successful people will be those who not only learn to incorporate new technologies and ways of thinking but are keeping an eye on the future and developing themselves accordingly along the way. This group includes those who are positioned to ride the wave of high-tech momentum, not just to survive the transformation but to thrive through it. They do not fear coworking or cohabitating with robots. Instead, they see the opportunity for future exploration and adventure. Such folks are eager to develop, implement, and allow advanced technological gadgets, gizmos, and systems to play their roles in making their lives and the lives of those around them more effective, efficient, comfortable, and profitable.

Changes are absolutely occurring—and they are happening at an exponential rate, slowly at first and then taking off. Although we do not yet know how far advanced technology will extend or how humanlike and independent robots will become, we know for sure that they are already affecting our lives and will probably continue to do so at a rapid pace. Regardless of the outcome, it is wise to begin making the necessary adjustments so that we are able to better keep pace with the changes and challenges as they arise. Acting on current trends and adapting to them accordingly can bring only benefits down the road. Failing to act, however, will ultimately lead to being unprepared for and at the mercy of whatever occurs.

If you have decided that you don't want to be left behind as the Age of Robots blossoms around you, then you need to make the decision to 1) get over the fear and frustration that it's happening, 2) get started by taking the actions necessary to put yourself more in line with the robot movement, and 3) understand who you are and develop your personal brand. Following are some of the ways that can be used to move into a more beneficial position in our robotic world.

GIVE YOURSELF A PSYCHOLOGICAL OVERHAUL

There is a popular quote by ancient Chinese philosopher Lao Tzu that says, "The journey of a thousand miles begins with a single step." That step in relation to successfully navigating exponential robotic growth is to give yourself a psychological overhaul by changing the way you think about advanced technology and yourself. It has been proven over and over again that negative thinking directly affects every aspect of our lives. If we think we can't—we can't. If we think we will fail—we will fail. The rise of robots can be extremely intimidating, especially if we think the end result will be world domination. Under such a negative mind-set, we tend to become frozen by fear and unable to take preparatory actions that lead to adjustment. The rapid pace at which robots are rising throughout society only compounds the fear, anxiety and defensive actions that thinking along those lines brings. A changing of the mind is needed.

So, how does one prepare for extreme changes that are accelerating at an exponential rate?

ABUNDANCE (HUMAN) AND
SCARCITY (ROBOT) THINKING

Through our culture and our education, we are trained to think in terms of scarcity. Scarcity simply means there is not enough for everyone. In contrast, abundance means there is more than enough for everyone. Because we're programmed to think in terms of scarcity, our alarm bells tend to go off when we hear talk of abundance. We come up with examples of what's not abundant, leap to the conclusion that a communist society is sharing and taking from hardworking people and giving to lazy people who do nothing, and so on. If you find yourself doing that, I ask that you put those notions aside and be open to learning a different way of looking at the world. Let me assure you that it won't lead to you living in a Soviet-like state. Instead, what I hope you see is a way of thinking that is more freeing and less stressful and that actually opens up economic opportunity.

First, let's understand scarcity thinking and where it comes from. You probably need look no further than your own childhood experience or that of your children to see kids fight over toys. Maybe you've heard the chorus of "Mine! Mine! Mine!" Maybe you didn't even want the toy until someone else showed an interest in it. Then suddenly you wanted it. You might even have been surrounded by toys you enjoyed, but now you wanted that one. It's very primitive, and it's programmed in us.

The entire discipline of economics is defined by the principle of allocating scarce resources. Where would economics be if there were too much of everything? This approach makes the math easy for optimizing existing ways of business and life. But here's what you also need to see: that is robot thinking. Will you outsmart a robot in mathematical operations? More important are the psychological impacts that prevent you from seeing opportunity, achieving success, and living a fulfilling life.

Scarcity thinking leads to closing your eyes to opportunities, thinking inside the box, and making a long list of excuses for why you don't have the life you dream about. It leads to blaming others for your condition. "I can't have this because _____ " (fill in the blank) or "If only this would happen, I'd be all set." This type of thinking is limiting and keeps you stuck.

Abundance thinking leads to a different thought process and set of questions. "I would like to have a _____; how can I do what's necessary to get it?" Abundance thinking is about possibilities. There is no limit in the way you achieve your goal. You let your mind work to figure out how to make it happen, even if you don't know just yet. Abundance thinking starts with the premise that whatever you want is possible; it's just a question of figuring out what needs to happen to get it. A good example is when President John F. Kennedy created a goal of putting a man on the moon within the decade. No one had done it. No one knew how to do it. There were many who thought it was impossible. They were partly right, because it wasn't possible with what they knew right then. A robot would have probably told them the same thing: it's not possible. There were too many pieces of the puzzle that hadn't been invented, and many weren't even known.

Abundance thinking is about possibilities. How can we make this happen? We need a computer that is not the size of a house but the size of a shoebox. Sounds outlandish, until you break it down and start to think about what needs to happen. What needs to be redesigned? The computer uses vacuum tubes; is there a way to achieve the same result using something different? As it happens, there is! This is the way successful people think too. It's also a very human and nonrobot way of thinking.

Use your human abundance thinking to successfully work with robots, algorithms, and other new technologies rather than compete with them using scarcity thinking.

LOOK FOR CHANGE AND HOW YOU CAN USE IT

Although our known methods for dealing with governmental, economic, and social upheaval may no longer serve us well, there are some principles that remain concrete, providing us anchors in stormy waters. Greek philosopher Heraclitus offered one of those principles many hundreds of years ago when he recorded an important observation: "The only thing that is constant is change." No matter how consistent, comfortable, and familiar times become, it is imperative that we fully expect changes to indeed come. The more we expect changes to happen, the less likely we will find ourselves blindsided by them and the more we will be in the business of preparing for the disruptions they ultimately bring.

EXPLORE

Fear is a common force in times of uncertainty, and when allowed to be blown out of proportion, it can be extremely debilitating—turning molehills into mountains. Because rapid, radical change upsets our cozy, habitual structure, we tend to embrace this powerful and often-negative emotion wholeheartedly. The problem with fear is that it prompts us to retreat further into the holes of our safe zones, where we simply wait for the horrific, inevitable outcome—biting our nails the entire time. Fear prevents us from making wise decisions or causes us to make unwise choices that compound problems.

People are creatures of habit, and when our habits or rules or systems are threatened by change, we tend to fear the outcome, resulting in our grasping ever tighter to those processes that are no longer beneficial. Thus, we often resist unconventional solutions while we cling to approaches that used to work but no longer are effective in the new paradigm. We therefore miss golden opportunities and make poor decisions based on outdated methods. Such actions only lead to more tumultuous disruptions that drive us deeper into fear and, ultimately, failure.

However, when we expect the arrival of major changes and have a decent understanding of how they will personally affect us, the fear of disruption subsides considerably. Instead of waiting apprehensively in our comfort zones, we are free to roam and explore options. Exploration is an extremely healthy activity anytime. It allows us to better see what's coming and to know what to do when shifts happen. It also presents opportunities to learn what works and what doesn't. We are able to get rid of those processes that no longer work and replace them with those that are more in sync with changing times—all before major disruption overtakes us.

THE POWER OF POSITIVE THINKING

It has also been well documented that positive thinking can transform the world around us for the better. If we think we will succeed, we draw to us those opportunities and resources that help us to face challenges and achieve success. Maintaining a positive mental attitude is vital in today's helter-skelter world, where new technological devices and systems are being pumped out on a near-daily basis. We must install and apply sound ways of thinking that give us a positive outlook to better see, anticipate, and master upcoming challenges.

Positive thinking doesn't mean being naive or unrealistic. It means that we take an optimistic view of what lies down the road. For example, it is a fact that robots are being developed that can perform a broader selection of jobs. A negative thinker will panic and become fearful or angry that her job is on the chopping block. A positive thinker, on the other hand, will agree that the field of robotics is expanding and taking over more jobs, but she will see it as an opportunity and make a career adjustment to fill a new position working with those robots, receiving an increase in income and more job security as a result.

Thinking optimistically will not only allow us to better prepare for coming changes and challenges, it will also make the journey from today and into the future more enjoyable. Happiness is one of life's most precious treasures, and it is well worth investing in by exerting the effort required to turn a fearful, negative attitude into one of courage, hope, and optimism. With the proper mind-set, we can do more than simply ride out the challenges of the robot age—we can enjoy the experience as it unfolds.

Existing lines of thinking must not only be overhauled to include an optimistic, positive outlook, they must also be trained to function exponentially, as was discussed earlier. It isn't sufficient to see progress according to traditional short-term and long-term goals. Exponential growth is occurring right here, right now, and a way of thinking must be grasped that matches this momentum. More than anything, it is our psychological and emotional adjustments that will need to be made and maintained in order to keep up with such rapid growth and change.

ADAPT TO CHALLENGES

Adaptation is a key ingredient in the formula for both surviving and thriving. When we are faced with events that challenge us and prod us out of our zones of comfort and familiarity, it is crucial that we remain resilient, make necessary adjustments, and bounce back to carry on—hopefully better than before. Adaptation is both necessary and powerful, especially when changes come to our personal, educational, social, and business environments. For the most part, humans are quite adaptable, being able to adjust when some new stimulus presses us to do so. As a matter of fact, the systems, laws, and protocols that we have grown up under and to which we are used to adhering are direct results of those people who preceded us learning to adapt to past challenges.

The ability to adapt is more valuable now than ever before, as the entire global community is in the process of major transformations on nearly every front. It's one thing to adapt to a single major challenge that occurs, but when challenges arise all around, it can be significantly more difficult to adjust. And in this age of disruptive, transformational technology, there can be simply too many adjustments to make all at the same time, which leaves us confused, drained of energy, fearful, and indecisive—all qualities that support failure and hinder success. We often believe that we are prepared for change and can handle the challenges that it brings but are left surprised and shell-shocked when reality hits. Under such inadequacy and pressure, we can quickly become lost in the swirl of challenges demanding our attention and be dragged down the drain of depression and defeat. This is why it is all the more important to understand who we are and what our personal brand is, which should remain constant. When we focus on the world external to us, there is always something more to chase after. When we know who we are, we can choose what will further enhance who we are and our unique promise of value—and ignore what doesn't.

When we are well centered, it is easier to roll with the robotic punches, persevere through the process of rising technology, and make adjustments that work to our benefit. In order to achieve this, we must remain patient and take small, necessary steps along the way so as not to be overcome by the magnitude of the challenges facing us. Staying focused on each challenge, and taking action to change with it, will lessen the anxiety and maximize the effort. It is much easier to maintain our footing through volatile terrain by taking small, calculated steps than to try and make long leaps in hopes of hastening the journey to more even and stable ground.

DON'T GET CAUGHT IN THE HYPE

It is even more critical to set goals in this age of robots and high technology. As scientist and futurist Roy Amara so accurately said, "We tend to overestimate the effect of a technology in the short run and underestimate the effect in the long run." This insight became known as Amara's Law, which demonstrates what has become known as a hype cycle, consisting of peak inflated expectations followed by a trough of disillusionment.[1]

Amara's Law and the hype cycle effectively characterize the atmosphere of the times in which we live. As new technologies are developed, they go from unknown to inescapable in a short period of time. Everyone is talking about it. During this time, some irrational levels of hope and fear can develop. Hopeful hype can lead to overinvestment, excessive risk-taking, and financial bubbles. Fear-based hype, which can often be seen in politics, similarly can lead to holding back too much, making excessively defensive moves, and missing out on opportunities. Hype is essentially setting unrealistic expectations that are bound to fail.

Following the hype is an even more dangerous time—a period of disillusionment. The new technology failed to live up to popular, unrealistic expectations, so it is dismissed. But the technology will continue to develop and move forward, regardless of whether you expect too much or too little of it. The period of disillusionment is dangerous for you because you dismiss the technology, but it will soon affect you, seemingly out of nowhere. It's OK to be excited about new tech; just be mindful of letting it cloud your judgment about whether you are looking at the current reality or a distant vision.

For example, superintelligent AI technology is thought to be available and running mainstream in the near future. We've been talking about robots, servant robots, lawyer robots, doctor robots, and so on. Does that mean we'll be inundated with them soon? What we know is that it's possible and that they will get better over time. How fast? It seems like they just came out of nowhere, when in reality they've been under development for quite some time. It could happen soon, or it could take decades—at least to develop the technology to a point where it is safe and constructive to use for our benefit. There is even the possibility that superintelligence may never be effectively developed. Such an atmosphere of unknowns makes it very difficult to make plans and set goals. Yet goals must be set in order to move forward, and you need to be ready for what does develop.

Technology itself can be used just as effectively as a positive for setting goals as it can be as a deterrent to them. Therefore, use existing and up-and-coming resources to help plan your strategies and set your goals. Don't be afraid to utilize technological means to assist with goal setting. Such resources can allow you to multiply your time, energy, and effectiveness toward achieving your goals.

TAKING CHARGE OF YOUR CAREER
AND FINANCIAL WELL-BEING

In the previous two chapters, you should have been able to get a better sense of yourself, the career you want, and how you will maintain your lifestyle. Now it's time to figure out how to make it happen. They say no battle plan survives first contact with the enemy; however, that's not a reason to not develop a plan. A plan allows you to clearly state your objectives so everyone on the team knows the objective and is empowered to do their part, by whatever means are necessary, to achieve that goal. Developing a plan will get you to think about problems before they arise and how you might handle them. If you know how to handle a problem when it arises, you can move past it faster. Most people spend more time and effort planning their vacation than they do planning their career.

START WITH THE END IN MIND,
BUT KNOW WHERE YOU ARE NOW

In chapter 8, you developed your vision, mission, and purpose statements and listed your motivating and burnout skills. Knowing where you are going and how you are going to get there is critical. Often left off the list, because it is taken for granted, is a solid understanding of where you are. Anyone who's used a map and compass to navigate to a destination is keenly aware of the need to know not only where you are going but where you are—and to have checkpoints along the way to ensure you are still heading in the right direction, especially if you need to take a detour to get around an obstacle.

The same is true with your career and your life. Rarely do people stop to assess where they really are today. There's a sense that maybe you aren't where you want to be and a list of reasons why you're not there. But what you need now is a solid assessment of where you are, no matter how painful or embarrassing it may feel. The important thing is that you don't want to stay there; you want to move on, and there is nothing more commendable than getting out of a tough, seemingly impossible situation and building your success.

It's time to create your status statement. Think of it as similar to your vision statement, but it's focused on the reality of here and now rather than the future. Perhaps your vision statement described you as the CEO of a Fortune 500 company. Your status statement would describe your current role working in the mail room of such a company, the tasks you do, which ones you like, which ones you don't like, and so on. You can describe your personal and financial life and anything else related to your vision.

TIME FOR A GAP ANALYSIS

At this point, you understand where you are today and where you want to be in the future. Now it's time to figure out what is needed to get you from where you are today to where you need to be in the future. In business, this is called a gap analysis—understanding the gap between where you are and where you want to be. Some people actually draw a map of different destinations as a strong visual aid. Others write out a list of points. The exercise below outlines the five key questions that you will need to answer. The answers to these incredibly important questions will provide you with a number of clear goals that can then serve as benchmarks for your plan.

1. What kind of education, training, or qualifications do I require?
2. What resources do I require?
3. What kind of connections do I require?
4. What kind of experience do I require?
5. How much time do I need to commit to achieving these goals?

Continuing with the brainstorming process, think about all the skills and experience you will need to gain in order to get from where you currently are to where you want to be. Do you need to gain managerial experience? Become more confident with public speaking? Sharpen sales skills? Learn more about your industry? Seek additional formal or informal education? What can you learn from your current job? What do you need to obtain another position within the same or a different company?

SET GOALS

The idea of setting goals can seem mind-boggling in the fast-paced environment in which we find ourselves, where nearly every segment of society around us is in constant flux. However, goals—both short- and long-term—are essential to a successful journey.

Use your gap analysis from above as the starting place for writing out your goals. It's important to write your goals and put them somewhere that you can see them daily. Having written goals significantly increases the likelihood you will work on them. It also provides clarity for what your goals are. Vague goals also rarely get achieved or move people forward.

When you write your goals down, do them in a way that allows you to change the order of them. Goal-setting tips include:

Identify any dependencies among goals: If one goal depends on the completion of another goal first, be sure to put them in the right order and make note of the dependency.

Order goals in terms of time priority: Which goals should you complete first? Dependencies are one aspect, but importance and achievability are also factors when deciding which goal to work on first.

Order goals in terms of importance: Which goals are the most important and least important? To the extent it is possible, put high-importance goals ahead of other goals in your prioritization.

Why are these goals important to you? Write a paragraph about why they are important. Include an emotional connection that will help you stay connected to your long-term goal. Keep it short.

When writing your goals, make sure they are SMART: specific, measurable, achievable, realistic, and time-bound. If you can, create smaller steps in between along the way—things that can be accomplished on a weekly basis. If you find yourself failing to achieve your goals, take stock and figure out why. Often this is an opportunity to test your values if you are failing because you are choosing other priorities over your goal. If your goals aren't aligned with your values, chances are you'll fail to achieve them.

FIND AN ACCOUNTABILITY PARTNER

People who are successful at achieving their goals have someone they work with who holds them accountable for doing what they said they would do. This can be a friend, a coach, or members of a group. Once you have made your choice, have a conversation about your goals and set up a weekly check-in (it can be as little as five minutes) to report on what you accomplished and what you intend to do the following week. If you failed to meet your previous goal, talk about what happened and how you might have changed the outcome. Make adjustments and move on with renewed enthusiasm toward meeting future goals.

CREATE YOUR STRATEGIC CAREER PLAN

All successful organizations create strategic long-term plans in order to make their vision a reality. They create such plans by doing a rigorous analysis of what the terrain is like between where they are now and where they want to be. This involves creating a written strategy that incorporates a solid understanding of who they are, the current situation, the desired end result, and everything they need to do to get where they want to be. It also involves identifying risks and providing a plan of action for what to do if those risks are encountered. The power of writing your strategic career plan is not that it won't change; it's in the clarity the foresight brings to you and the increased adaptability that comes with it. And by now you've created all of this information for yourself, so it's time to see how it all works together for you!

Businesses use a strategic plan to set direction and priorities to ensure that all stakeholders (which includes owners, employees, investors, and others) are effectively working toward the same goals. A strong strategy is detailed enough to provide discipline in a world full of distractions, yet flexible enough to adapt to a changing environment. A solid strategic plan includes the company's destination, how it will arrive there, and the metrics to distinguish whether the provided methods were successful.

Following are the common elements of a strategic plan:

Values statement
Vision, mission, and purpose statements
Unique promise of value/branding/competitive advantage
Strengths, weaknesses, opportunities, and threats
A list of goals, objectives, and accountable partners

If you've been working along during the last few chapters, you should have all of the elements of a strategic plan for yourself. Now just put them all together in a single document. Read it through. Does everything fit together? The first time through, it may not all seem to work together, so go back and connect any missing dots, and add anything that seems to be lacking. You don't have to spend a lot of time polishing and perfecting it, though. A plan is a useful tool, but it'll need to be adapted and changed as you go along. Perfect it to the point where you know what you need to be doing. Taking action is far more important than having a perfect plan. Progress, not perfection.

The strategic planning process works well for billion-dollar companies, and it works well for You Inc., the business of you. As you probably noticed, you've already accomplished most of what is included in a good strategic plan because you should already have your values, vision, mission, and purpose statements; know your motivating and burnout skills; and be developing your personal branding statement. However, if you want to bring the vision that you have created for You Inc. to life, you need to do the analysis of what it will take to get from where you are now to where you want to go. When you plan and act strategically, you look ahead and plan for contingencies and for times when events occur that are outside of your control. When you think strategically, you look beyond your next move to your second, third, or fourth.

What exactly are you evaluating? How do you know if you are on the right career path? How do you even know what career path to be on? While it is easy to say you should evaluate your progress, most of us struggle to know what path we want to be on, what we need to do to get where we want to be, and what good progress actually is. When you have a strategic career plan or life plan, you have a road map against which you can evaluate your progress. If you don't have a strategic career plan, it will be difficult to say whether you are hitting your goals or not. In fact, it becomes very easy to just let your career go with the flow and end up nowhere in particular. I call this having a "career by accident." If you are not intentional in your career direction and path, you have no way to determine if you are on the right path or if you are achieving the success you want.

The fact is that many people change their career plans over the course of their lifetime. Your vision for You Inc. might change in five or ten years because you learn something new, discover a passion you never knew you had, change the way you see the world in a fundamental way, or are forced to adjust due to outside elements. If that happens to you, simply make the necessary adjustments to your strategic career plan and continue. It is important not to make the common mistake of becoming locked into an idea of what you "want" to be or do based on an adopted vision of someone else—such as becoming what your parents want for you. When you know what you want and develop a strategic career plan to guide you to your destination, you will have the drive and desire to get there—even if the course changes along the way.

With your strategic career plan in hand, you now have the power to act on and follow it toward your desired goal. Be sure to get started right away while your energy and fervor are strong. Waiting for circumstances to be just right will only delay your success. You can always make adjustments to your course along the way, but it is important to get the journey started. And remember, this is *your* plan—you can change the destination at any point!

STAY MOTIVATED: LET'S GO!

Even if you have great ideas and a strong vision of what they can accomplish, if you aren't motivated to see plans through to the end, they simply will not manifest into reality. The business world is littered with dead ideas that once shone brightly with hope. However, if you are passionate about what you are doing, that energy will not only get you going, it will also help carry you through until you meet your goal. Passion is particularly helpful to drive motivation when the dark times come (and they will come), keeping you going when everything around you screams, "The end is nigh!"

Motivation is a powerful quality that allows 'preneurs to build great companies out of what seems like thin air. It is the deep-down desire to accomplish the vision come hell or high water. 'Preneurs are passionate, and they are motivated. They consistently maintain a behavior that is disciplined and goal oriented, allowing them to continue making strides toward success. Motivation is a quality that turns passivity into leadership, fear into faith, and indecisiveness into decision-making. It gives those who hold it the ability to look doubt in the eye, push it aside, and move onward, overcoming the obstacles it tries to cast in their path.

Time is something that every single one of us has in common. We all have twenty-four hours in a day, seven days in a week, four weeks in a month, and twelve months in a year to fulfill our dreams and accomplish our visions. The difference between 'preneurs and the rest is that they utilize the allotment of time in a more organized and efficient way. It may seem like the makers and shakers have more time to accomplish great things, but the truth is that they are simply passionate about their visions and therefore motivated to fulfill what they set out to do, and they use that energy to set priorities. Instead of sleeping in until eight, they are up earlier in order to get a jump on the day. Instead of turning on the television or streaming a movie, they mull over problems, proposals, and plans. Instead of heading out for social events, they remain in their offices, shops, and plants seeking ways to develop, improve, and implement their ideas. Motivation provides 'preneurs with the self-discipline necessary to turn down temptations, fight through fatigue, overcome obstacles, and get things done.

Motivation also gives 'preneurs the courage to make their own rules, and even though they are surrounded by both internal and external groups and forces, they continue to march to their own drum because they know what it takes to reach their goals. Speaking of goals, motivation causes 'preneurs to push the limit. They don't set simple, comfortable goals. No, they set goals that are aggressive, and then they attain them, accomplishing desired results much faster than expected—which is another way in which they give the illusion of having more time. And regardless of how long or difficult the journey, 'preneurs never lose sight of their destination. Their consistency comes from setting obtainable short-term goals that are key in reaching longer-term goals, but those targets sitting in the distance challenge them to venture outside their comfort zones to where miracles are created and dreams come true.

FLEXIBILITY: BEND WITH THE WIND

Anyone who wants to make it in the business world must be flexible. Even in the most stable times, business owners, investors, and others must deal with the fluid predictions of financial wizards, changes in market landscapes, fluctuations in pay scales, tax laws, product and supply prices, finicky fads of customer bases, and many other variables that may transform abruptly or over time. Flexibility is a must-have in today's turbulent business world, where all of the abovementioned topics and more tend to change sooner or later, since market indicators, currency rates, government regulations, and a thousand other things remain in a near-constant state of flux.

In all of this chaotic change, how are so many 'preneurs able to rise to the occasion and flourish? They remain flexible, and that flexibility comes from practicing a number of key traits. First of all, 'preneurs are multifaceted in how they view their surroundings, constantly monitoring and analyzing the business climate that surrounds their particular element as well as those that directly affect it. During this process, they create alternative short-term plans that they can implement at the drop of a hat should something change. A plan B strategy is also created to fall back on so that little, if any, momentum is lost when changes actually occur that are significant enough to halt the original plan.

Next, 'preneurs are optimistic but not gullible. They do not assume things will automatically work out to their benefit. Rather, they remain vigilant and ready to make the necessary decisions and moves that progress their plan. 'Preneurs don't presume to know how long the process will be to develop their idea to fruition, or how tough the competition will be, or whether something similar will come along to challenge their idea. Instead, they press ahead toward meeting short-term goals, and should any hurdles arise during the journey, they simply weather the storm and keep things moving toward the long-term target.

'Preneurs are also flexible when it comes to resources. They know how to utilize their gains and push their ideas in fat times, and how to tighten their belts and trim expenditures in thin times. Optimism about reaching their goals is always at the forefront, shining like a beacon on the dark waters ahead, and flexibility acts like a rudder that turns their efforts instantly to avoid obstacles. How are obstacles identified? By constant testing of the immediate waters and unfaltering vision that scans the horizon. These actions, coupled with the flexibility to make necessary moves at any time during the process, lead to a successful outcome.

If you're not already thinking and acting like a 'preneur, there is no better time to start than right now. The time to start learning skills and adopting a new mind-set is long before you find human resources holding the door open for you on your way out. As personal branding expert Dan Schawbel said, "You need to find a way to make a unique contribution, add value, and stand out. That's the only way to survive."[2]

ACCELERATE RESPONSE SPEED

Exponential thinking is not only relevant for accurately assessing patterns and trends and excelling at educational development, it is also a critical element in making key decisions under the pressures of a rapidly transforming technological landscape. A flurry of new developments in robotics and artificial intelligence is creating vast changes that must be identified and adjusted to in order to succeed personally, socially, and corporately. As we've discussed, those who lag behind can be quickly overwhelmed, and the longer this resistance, inactivity, or procrastination occurs, the further behind you fall. In a time of exponential growth, that can spell complete failure on many fronts.

The ability to recognize needs or opportunities and make decisions toward meeting them is known as response time. Rapid decisions lead to equally rapid actions that address pressing issues. A fast response time is yet another key element that makes entrepreneurs successful, and it is just as vital for making adjustments in our transforming world. During previous centuries and decades, when the cogs of progress moved at a slower pace, there was usually ample time to research, think things through, come to sound decisions, and take action concerning pressing issues. That is definitely not the case today. Critical changes are occurring monthly, weekly, and even daily that can drastically affect lives on a variety of levels. In order to meet the challenges they bring, response time must be accelerated. We are increasingly offered smaller and smaller windows of time to make vital decisions, yet if we fail to do so, we can quickly join the resisters, do-nothings and procrastinators in the expanding mire pit out of which it becomes increasingly difficult to climb.

A fast response speed for making critical decisions is amplified by the characteristics of the 'preneur discussed in the previous chapter, which is why such characteristics—and the people possessing them—are so critical. When you are eager, excited, fearless, insightful, and calculated, you are able to forge ahead and keep pace in a world where an upward-spiraling tornado of changes is occurring. Accelerating your decision-making response time increases a number of positives. It opens opportunities for employment as well as investment, and it helps keep you a step ahead of competitors. It also helps you meet customer needs in a timely manner—useful because customers can be finicky and trendy in nature, ready to change and move elsewhere on a whim. You've probably noticed that people are becoming less patient as the age of technology races forward, so making fast decisions to meet customer needs and demands has become a critical factor in today's business world.

HONE BUSINESS SKILLS: GET 'ER DONE!

Having honed skills hovers at the top of the list when it comes to building a successful career. Even if you possess the knowledge of how to achieve a certain project, if that knowledge is not backed by the skills needed to apply it, it does little to achieve the task. Take, for example, a comparison between an airplane mechanic who has just graduated from trade school and one who has been serving in the air force with his head tucked in the fuselage of fighter jets for twenty years. Who would you want working on your state-of-the-art private jet? Well, the answer might depend on their knowledge and skills with the new technology. Someone with twenty years of experience working with outdated technology may be less valuable or trusted than someone with two years of working with new technology. If you are the person with twenty years of experience working with old tech, use those years of experience for the wisdom they provide in finding and solving problems, but work on your skills with the technology too.

Honed skills are just as important for 'preneurs to possess. Eagerness, vision, and even base knowledge may catch the attention of others, but it is the possession of skills gleaned from real experience that compels people to get behind you and your idea (remember persuasion?). You therefore need to identify the main skills required by 'preneurs and hone them to the highest degree possible to produce the results that 'preneurs achieve in the real world. So what are those skills? Let's consider some of the most prominent.

'Preneurs have the ability to look at issues from a variety of angles. This gives them a unique view of how specific moves and actions may work to their advantage. They have an uncanny sense of how markets fluctuate and regroup, which allows them to pinpoint rising trends with acute accuracy. This sense of navigating through turbulent times only strengthens with practice and experience. At the same time, 'preneurs maintain a laser focus on the inner workings of their business, which works to keep everyone on point and moving toward set goals.

'Preneurs are known for being visionaries who create and identify value while letting managers worry about maintaining internal organization. It can often be difficult for 'preneur types to fill managerial roles, but the benefits of honing such skills are beneficial. Entrepreneurs from Henry Ford to Mark Zuckerberg have successfully made the transition, and the results are obvious. It serves a 'preneur well to hone her management skills so she has more control over her projects—at least in a guiding sense, if not outright oversight.

There are many other skills that can be honed that will faithfully serve the entrepreneur, many of which are geared more toward individual and personal goals. Simply identify what skills will best carry you to your destination and begin practicing them until they become second nature. The rewards are well worth the effort.

ENDURE RISK: FAILURES ARE STEPPING-STONES!

Every single business venture has some degree of risk associated with it because there is a long and ever-changing list of events, shortcomings, and unexpected issues that can occur to stop progress on the endeavor. Investment capital could dry up or not materialize as planned; the product or service may not catch the market share you anticipated; regulations involving the industry could dramatically change; there might be failure to draw key talent and support; or you could lose those you are depending on to develop, produce, or market your product or service. And those are just some of the major missiles that could sink your project.

Since risk is a necessary element in the business success formula, it should be expected. As American journalist and humorist Verni Robert Quillen so aptly wrote in 1924, "Progress always involves risk. You can't steal second and keep one foot on first base." There is some level of risk involved both during start-up and in every step of the progressive process. Unfortunately, a large number of those who start businesses are unstructured risk-takers who go about it with a "make it or break it" mentality that somehow expects doors to open and needs to be met magically along the way. The problem with such an approach is that collapse occurs early in the process more often than success, largely because of this roll-of-the-dice attitude. According to the US Bureau of Labor Statistics, nearly 80 percent of businesses end up surviving their start-up year, but the rate of survivability drops to around 66 percent the second year and falls to about 50 percent by the fifth year of operation. Out of all businesses, only roughly one-third will hit the ten-year benchmark.[3]

Of course, a small number of the extreme risk-takers make it simply due to luck, just like a new gambler will occasionally win a big pot, but a professional gambler will confess that consistent wins are due not to luck but to carefully calculating the odds when making bets. The majority of those business ventures that weather the storms of the first few critical years and become established are the 'preneur types. The difference between the average person pursuing a business start-up and a successful 'preneur is that the latter tends to take calculated risks—just like the professional gambler. 'Preneurs consider all of the available data and create a plan of execution that reduces risk. At every step, they cautiously apply their model, which not only prevents big losses but expands the impact of their investment funds. Along the way, they are constantly recalculating results and searching for less expensive, faster, and more effective ways of operating.

The taking and successful maneuvering through risky terrain creates leaders that others want to support, invest in, and follow. These 'preneurs turn raw material into low-risk opportunities, which are then used to propel momentum and inspire positive change. Such leaders are what it takes to turn turbulent times into a more organized and navigable landscape where exponential growth can be harnessed and used to make the world a better place. Calculated risk-takers also use their innovation to make their ideas common products and services. They push the envelope of what can be to discover new possibilities and methods for dealing with changing times.

No one is failure free. At times, even 'preneurs encounter failure, but instead of letting such events drive them into despair, they use them as stepping-stones to achieve even greater success. The list of entrepreneurs who have failed (many of them miserably and often) but have risen to achieve great things is a long one. Their testimonies trumpet the truth that lessons gleaned from failures are powerful elements in the process. Taking calculated risks and then failing offers the chance to analyze those situations, learn from what went wrong, and then use the new knowledge to modify the process so that subsequent attempts are more successful.

PERSEVERANCE

Another quality highly sought after in today's business world is perseverance, or as Greg S. Reid referred to it, "stickability."[4] This is a quality held by every successful leader, which gives them the power to turn lofty ideas into reality and achieve critical goals. Perseverance requires laser focus and strict discipline to resist opposition and find ways to overcome problematic issues. Those who persevere usually embrace other strong qualities such as being responsible and accountable, which also give them a good work ethic. They tend to have or develop defined goals and priorities because they walk the line until successful steps are revealed. They have faith in their goal or vision, even if the path itself isn't clear. When they run into obstacles, they get help and advice from experts and mentors to show them the way.

If you feel discouraged about your goals, but you still believe in them, keep in mind the story of R. U. Darby. His story is recounted in *Think and Grow Rich* and *Three Feet from Gold*.[5] Darby was a gold prospector who found a vein and continued along it until the vein seemingly ran out. Frustrated, he gave up, sold his equipment, and moved back home. The prospector who bought Darby's equipment asked an expert, who said that the geology had shifted due to a fault line. All he needed to do was move three feet further and pick up the vein again. Sure enough, it was there!

R. U. Darby gave up just three feet from gold. As we move further into the fourth industrial revolution with opportunity all around us, will you do the preparation you need to be successful? Will you recognize opportunity when you see it? Will you be ready to seize that opportunity? Will you rely on the counsel of experts to help you get there? Or will you join R. U. Darby and give up before your time?

It's your career. Your life. You have the opportunity to achieve success however you want to define it, unlike ever before. While technology continues to advance, the same time-tested disruptions to your career and financial well-being are still with us. Economic business cycles. Corporate mismanagement. Political change that favors and disfavors certain industries and businesses. Corporate restructuring. In *The Robot in The Next Cubicle*, I've shared just a small number of the changes that are happening. It is my hope that you have a better understanding of the breadth and depth of the change that is coming in the next few years and that you use this time to develop your personal strategy to succeed. Our elected officials are turning a blind eye to the changes, changes that CEOs say are coming from both technologies they are developing and personnel decisions they are thinking about. Innovation Centers and start-ups are actively working on developing disruptive products and business models. But no matter how much optimism they have for job creation, those new jobs will only help you if you have the skills needed for those jobs and are willing and able to move to where they are.

Your future is in your hands. Know what makes you valuable. Continue to learn and develop yourself so you can continue to add value throughout your lifetime. The time to future-proof yourself is now. When you've done the preparation and understand the new business and technology landscapes, you can confidently pursue the opportunities you want on your terms. Define your success and make it happen!

FINAL THOUGHTS

The number of scientific breakthroughs and business innovations that are occurring is staggering. It is critical to note, however, that as exciting as groundbreaking discoveries are, it takes time to see their commercial impact. This gives the illusion that our jobs and our savings are safer than they are. Today's breakthroughs, some of which have been mentioned in the course of the book, will become game changers that disrupt business in the years to come. Keep in mind how exponential technologies at first grow more slowly than linear expectations but then are suddenly everywhere. Companies rise and fall with these shifts, and along with them the financial well-being of their employees. Disruption, whether from technology, business cycles, political change, or bad business decisions, happens fast.

When corporate executives warn us that jobs will be cut, are we listening? Companies like AT&T® have put their workforce on notice to learn new skills or else find their future to be "limited." AT&T alone could find itself in need of one-third fewer employees in just a couple of years.[1] John Cryan, CEO of Deutsche Bank®, has hinted that his company could see its work force of ninety-seven thousand employees cut in half in the coming years.[2] And these companies are not alone in talking about the future of their employees. Make it a point to listen to what the CEO of the company you work for is saying about the direction of the company.[3] Your ability to control your future may depend on it.

As I write this in the closing days of 2017, there's been an explosion of interest in cryptocurrencies, especially Bitcoin, Ethereum®, and Litecoin®. No doubt they, and others, will continue to grow. What's being missed by most people looking for a high return on investment is the blockchain projects these coins back. The objective of the companies behind these coins is to disrupt existing businesses—basically, take their business away. Some of these companies, along with their coins, will succeed wildly and meet their purpose. Most will likely fail for the usual reasons start-ups fail. Existing business won't sit still either. But rather than compete and build using new technologies, they will fail to use their competitive advantages through lobbying for regulation.

Artificial intelligence, robotics, and analytics will continue to grow and develop. There is a lot of hype about the potential of each, and we tend to think about the amazing machine intelligence we see in the movies. There are those who point to those intelligences and suggest we are far from achieving them. They may be right; however, fully realizing the possibilities of AI may not be necessary in order to replace you at work or put your company out of business. We can already see how relatively simple bots are curtailing the need for Wall Street traders, investment fund managers, and financial advisors. While you can look at a robot, like Sophia, and feel comfortable that it won't take your job, that's the wrong place to be looking. It's what's not flashing in the news that you need to be looking out for.

It can be quite unsettling to look over and see a robot coworker in the next cubicle, or know that one will most likely take that position in the very near future. It is even more disturbing to think about how your job could one day be done by a mechanical employee that does a faster, better job at a cheaper rate and without the need for costly benefits. We are rapidly heading into the "great unknown" as robots and AI systems arc created, developed, and tweaked to a higher and higher performance ability. Robots absolutely present some very real threats to large segments of human society, but they just as equally provide some major opportunities for extreme change that can be very positive indeed.

The news-catching headlines of artificial intelligence are also easy to dismiss as curious parlor tricks. Google's AlphaGo® robot offers an interesting example. It has a very limited purpose, so by itself it's not very threatening, except perhaps to human pride. However, it's the "thinking" process that can be generalized and applied to other issues that truly matters. As a reminder, in 2016, Google stunned the world when AlphaGo beat that year's world champion Go player, Lee Sedol.[4] The original AlphaGo system was designed by Google's DeepMind® division and was programmed with not only game rules but also human expert input of around thirty million moves. Reigning champ Sedol was only able to win a single game out of five against the robot, which could accurately anticipate his moves 57 percent of the time. DeepMind developers went right to work on the second AI player, dubbed AlphaGo Zero®. The difference this time was that AlphaGo Zero was only programmed with the game rules and allowed to learn by itself as it went, without human intervention. The robot began making random moves, and eventually its learning became exponential in nature. After three days, the new robot reached the point that surpassed the performance of the original AlphaGo against Lee Sedol. At the twenty-one-day mark, it beat an online updated version of the game (AlphaGo Master®) that pitted the robot against sixty game experts. It ended with a win-loss record of 89 to 11, which included winning three out of three games against 2017 world champ Ke Jie. Forty days from the point AlphaGo Zero began self-learning the game, it was proclaimed to be the best player in the history of the world—all without using any human intervention or historical data of game strategy.[5] This is a glimpse into the future of AI systems and their potential.

How it all turns out with the increased momentum of robots and artificial intelligence is still very much up in the air. What we do know is that it's coming, and we have to make the choice to deny it, fear it, or prepare for it. Since the first two options would leave us totally caught off guard and lacking ability to adequately adjust to the changes, it is best to do all we can to prepare. However, the gap of opportunity is shrinking and squeezing out many people who are not adequately prepared for such rapid and extreme changes. The time to act is now. If you remain vigilant and flexible, you will be able to make the adjustments necessary to successfully navigate the multiple and rapid changes of our growing environment of robots. A vast number of opportunities are arising that could be quite enjoyable and lucrative for those who identify, prepare for, and take advantage of them.

The key to riding the wave of success in the Age of Robots is to harness their amazing power while at the same time fine-tuning your human knowledge and talent base and your sense of who you are. This *is* the wave of the future, and it requires the development and implementation of the principles found within these pages. I have written this book as a guide to help you understand the nature of what is happening all around us and provide you with some basic knowledge of how to vastly improve your chances of not only surviving the coming Age of Robots but actually enjoying and benefiting from the robot in the next cubicle.

One of the great lessons I learned modeling and forecasting financial risk and outcomes is that there are two components to managing future risk. The first is to predict what you can and prepare for it (predict how many loans will default and set aside money to cover the losses). The second acknowledges that you can't predict everything, and therefore you must manage the risks and outcome. For example, you can't predict whether an earthquake will destroy a building, and you couldn't charge a high enough premium to cover such a catastrophic loss. What you can do is manage the risk—require stronger building materials, reinforcements, and so on to help reduce the damage and risk of injury. Similarly, you can't possibly know all the risks to your career. Some you may be able to see and predict; others you may not. Where possible, anticipate changes, learn, and adapt. Where that's not possible, prepare yourself for a job loss, extended unemployment, and a different career direction.

An important aspect to keep in mind is that when it's all said and done, we are all in this together. Supporting one another provides a powerful advantage over those lone wolf types who seek to bury everyone in their paths and make it on their own. Pulling together will become even more valuable as advanced technology brings about the abrupt changes and associate disruptions that we have discussed in this book. If you are looking to be a part of a support group that keeps a close eye on the topics in this book, consider joining the conversations on http://www.futureproofingyou.com, where you can enter open discussions to share information, ideas, and opportunities to help each other navigate our disruptive future.

NOTES

INTRODUCTION

1. "Fortune 500: A Database of 50 Years of FORTUNE's List of America's Largest Corporations," *Fortune*, http://archive.fortune.com/magazines/fortune/fortune500_archive/full/1965/ (accessed November 5, 2017).

2. "Current World Population," Worldometers, http://www.worldometers.info/world-population (accessed November 15, 2017).

3. Max Roser and Esteban Ortiz-Ospina, "Future World Population Growth," Our World in Data, last updated April 2017, https://ourworldindata.org/future-world-population-growth (accessed October 15, 2017).

4. Or at least, this saying is widely attributed to him. See Wikiquote, s.v. "Alexander Graham Bell," last modified April 9, 2018, 13:39, https://en.wikiquote.org/wiki/Alexander_Graham_Bell.

CHAPTER 1. THE FOURTH INDUSTRIAL REVOLUTION: A PROGRESSION TOWARD PERFECTION

1. Richard Tames, *Documents of the Industrial Revolution 1750–1850: Select Economic and Social Documents for Sixth Forms* (Abingdon, UK: Routledge, 2013), pp. 111–12.

2. Ibid.

3. Klaus Schwab, *The Fourth Industrial Revolution* (New York: Crown Business, 2017), pp. iii–iv.

4. Ozonnia Ojielo, "Access to Technology Can Help Prevent Violent Conflicts," United Nations Development Programme, August 7, 2012, http://www.undp.org/content/undp/en/home/ourperspective/ourperspectivearticles/2012/08/07/access-to-technology-can-help-prevent-violent-conflicts.html (accessed March 10, 2018).

5. "What Is Genomic Medicine?," National Human Genome Research Institute, last modified November 28, 2017, https://www.genome.gov/27552451/what-is-genomic-medicine/ (accessed March 20, 2018).

6. Josh Constine, "Facebook Is Building Brain-Computer Interfaces for Typing and Skin-Hearing," *TechCrunch*, April 19, 2017, https://techcrunch.com/2017/04/19/facebook-brain-interface (accessed March 20, 2018).

7. Nick Statt, "Elon Musk Launches Neuralink, a Venture to Merge the Human Brain with AI," *Verge*, March 27, 2017, https://www.theverge.com/2017/3/27/15077864/elon-musk-neuralink-brain-computer-interface-ai-cyborgs (accessed March 20, 2018).

8. Andrew J. Hawkins, "Elon Musk Thinks Humans Need to Become Cyborgs or Risk Irrelevance," *Verge*, February 13, 2017, https://www.theverge.com/2017/2/13/14597434/elon-musk-human-machine-symbiosis-self-driving-cars (accessed March 20, 2018).

9. Ameer Rosic, "What Is Cryptocurrency: Everything You Need to Know {Ultimate Guide}," BlockGeeks, March 2017, https://blockgeeks.com/guides/what-is-cryptocurrency (accessed October 12, 2017).

10. Panos Mourdoukoutas, "Four Facts about Bitcoin," *Forbes*, September 18, 2017, https://www.forbes.com/sites/panosmourdoukoutas/2017/09/18/four-facts-about-bitcoin/#10a0f5225abf (accessed March 20, 2018).

11. Christopher J. Goodman and Steven M. Mance, *Employment Loss and the 2007–2009 Recession: An Overview* (Washington, DC: US Bureau of Labor Statistics, April 2011), table 2, https://www.bls.gov/opub/mlr/2011/article/employment-loss-and-the-2007-09-recession-an-overview.htm (accessed March 20, 2018).

12. Wade Roush, "Hamburgers, Coffee, Guitars, and Cars: A Report from Lemnos Labs," Xconomy, June 12, 2012, http://www.xconomy.com/san-francisco/2012/06/12/hamburgers-coffee-guitars-and-cars-a-report-from-lemnos-labs/# (accessed March 20, 2018).

13. Melia Robinson, "This Robot-Powered Restaurant Could Put Fast Food Workers Out of a Job," *Business Insider*, June 30, 2016, http://www.businessinsider.com/momentum-machines-is-hiring-2016-6 (accessed March 20, 2018).

14. Ibid.

15. Jonathan Amos, "'Robot Chef' Aimed at Home Kitchen," BBC,

April 14, 2015, http://www.bbc.com/news/science-environment-32282131 (accessed March 20, 2018).

16. Fred Lambert, "Elon Musk Goes on a 'Machines Building Machines' Rant about the Future of Manufacturing," Electrek, June 1, 2016, https:// electrek.co/2016/06/01/elon-musk-machines-making-machines-rant-about -tesla-manufacturing/ (accessed March 20, 2018).

17. "Elon Musk Quotes (various)" Elon Musk News, https://elonmusk news.org (accessed October 30, 2017).

18. "About OpenAI," OpenAI, https://openai com/about/#mission (accessed April 10, 2018).

19. Sam Wong, "Google Translate AI Invents Its Own Language to Translate With," *New Scientist*, November 30, 2016, https://www.newscientist .com/article/2114748-google-translate-ai-invents-its-own-language-to -translate-with (accessed March 20, 2018).

20. Kevin Maney, "How Facebook's AI Bots Learned Their Own Language and How to Lie," *Newsweek*, August 5, 2017, http://www.newsweek .com/2017/08/18/ai-facebook-artificial-intelligence-machine-learning-robots -robotics-646944.html (accessed March 20, 2018).

21. Valerie Hamilton, "No Farm Workers? How about a Robot," Public Radio International, June 2, 2017, https://www.pri.org/stories/2017-06-02/ no-farm-workers-how-about-robot (accessed March 20, 2018).

22. Jacob Kastrenakes, "A Cartoon Mark Zuckerberg Toured Hurricane-Struck Puerto Rico in Virtual Reality," *Verge*, October 9, 2017, https://www .theverge.com/2017/10/9/16450346/zuckerberg-facebook-spaces-puerto -rico-virtual-reality-hurricane (accessed March 20, 2018).

23. Harold L. Sirkin, Michael Zinser, and Justin Rose, "The Robotics Revolution: The Next Great Leap in Manufacturing," BCG Perspectives, September 23, 2015, https://www.bcg.com/publications/2015/lean -manufacturing-innovation-robotics-revolution-next-great-leap-manufacturing .aspx (accessed March 20, 2018).

24. Tom Green, "Investing in Robotics 2016: What's Happening and Why," *Robotics Business Review*, April 8, 2016, https://www.robotics businessreview.com/manufacturing/investing_in_robotics_2016_whats _happening_why/ (accessed March 20, 2018).

25. Patrick Thibodeau, "China Builds World's Fastest Supercomputer without US Chips," *Computerworld*, June 20, 2016, https://www.computer world.com/article/3085483/high-performance-computing/china-builds -world-s-fastest-supercomputer-without-u-s-chips.html (accessed March 20, 2018).

26. "China's Xi Pushes Advanced Technology for Military," Reuters, March 12, 2017, http://www.reuters.com/article/us-china-parliament -defence/chinas-xi-pushes-advanced-technology-for-military-idUSKBN 16K02V (accessed March 20, 2018).

27. Cate Cadell and Adam Jourdan, "China Aims to Become World Leader in AI, Challenges US Dominance," Reuters, July 20, 2017, https:// www.reuters.com/article/us-china-ai/china-aims-to-become-world-leader-in -ai-challenges-u-s-dominance-idUSKBN1A5103 (accessed March 20, 2018).

28. Ian Bogost, "Why Zuckerberg and Musk Are Fighting about the Robot Future," *Atlantic*, July 27, 2017, https://www.theatlantic.com/ technology/archive/2017/07/musk-vs-zuck/535077/ (accessed March 20, 2018).

29. Larry Boyer, "Jobs, AI, and Automation: What You Need to Know," *Forbes*, November 3, 2017, https://www.forbes.com/sites/forbescoaches council/2017/11/03/jobs-ai-and-automation-what-you-need-to-know/ #19890848cb5c (accessed March 20, 2018).

CHAPTER 2. THE OBSESSION WITH AUTOMATION

1. Don Stewart, "When Did Moses Write, or Compile, the Book of Genesis?" Blue Letter Bible, https://www.blueletterbible.org/faq/don _stewart/don_stewart_678.cfm (accessed October 1, 2017).

2. Homer, *Iliad* 18.371.

3. Only in *Argonautica* 4 and in *Eustathius*, according to H. de la Ville de Mirmont, *Apollonios de Rhodes: les Argonautiques: traduction française suivie de notes critiques* (Paris and Bordeaux), 1892:402, noted in J. Douglas Bruce, "Human Automata in Classical Tradition and Mediaeval Romance," *Modern Philology* 10, no. 4 (April 1913): 511–26; 513 and note.

4. "The Myth of Pygmalion and Galatea," Greek Myths & Greek Mythology, http://www.greekmyths-greekmythology.com/myth-of -pygmalion-and-galatea (accessed October 1, 2017).

5. "Automotoncs," Theoi, http://www.theoi.com/Ther/Automotones .html (accessed October 1, 2017).

6. Joseph Needham, *History of Scientific Thought*, Science and Civilization in China 2 (Cambridge: Cambridge University Press, 1956), p. 53.

7. Linda Safran, *Heaven on Earth: Art and the Church in Byzantium* (Pittsburgh: Penn State Press, 1998), p. 30.

8. "The Alchemical Quest," *Occult Science & Philosophy in the Renaissance* (blog), Louisiana State University Library, http://exhibitions .blogs.lib.lsu.edu/?p=1257&page=5 (accessed October 1, 2017).

9. Gunalan Nadarajan, "Islamic Automation: Al-Jazari's Book of Knowledge of Ingenious Mechanical Devices," Muslim Heritage, http://www .muslimheritage.com/article/islamic-automation-al-jazari%E2%80%99s-book -knowledge-ingenious-mechanical-devices (accessed October 1, 2017).

10. Teun Koetsier, "Mechanism and Machine Theory," *Elsevier* 36, no. 5 (2001): 589–603.

11. Charles B. Fowler, "The Museum of Music: A History of Mechanical Instruments," *Music Educators Journal* 54, no. 2 (1967): 45–49.

12. Mark Rosheim, *Robot Evolution: The Development of Anthrobotics* (Hoboken, NJ: Wiley, 1994), p. 36.

13. Sylvia Landsberg, *The Medieval Garden* (London: British Museum Press, 1995), p. 22.

14. Joseph Needham, *Physics and Physical Technology*, vol. 4, *Science and Civilization in China* (Cambridge: Cambridge University Press, 1971), pp. 133, 508.

15. Mark Rosheim, *Leonardo's Lost Robots* (New York: Springer, 2006), p. 69.

16. Elizabeth King, "Clockwork Prayer: A Sixteenth-Century Mechanical Monk," *Blackbird: An Online Journal of Literature and the Arts* 1, no. 1 (2002), https://blackbird.vcu.edu/v1n1/nonfiction/king_e/prayer_introduction.htm (accessed March 20, 2018).

17. Edgar Allan Poe, *Edgar Allan Poe: Essays and Reviews*, ed. Gary Richard Thompson (New York: Library of America, 1984), p. 1253.

18. Rodney Allen Brooks, *Flesh and Machines: How Robots Will Change Us* (New York: Knopf Doubleday, 2002), pp. 14–15.

19. "Maillardet's Automaton," Franklin Institute, https://www.fi.edu/history-resources/automaton (accessed October 1, 2017).

20. "The Silver Swan," Bowes Museum, http://thebowesmuseum.org.uk/Collections/Explore-The-Collection/The-Silver-Swan (accessed October 1, 2017).

21. "Musical Automation," Waddesdon Manor, https://waddesdon.org.uk/the-collection/item/?id=39 (accessed October 1, 2017).

22. Ivan Karp and Corinne A. Kratz, "Reflections on the Fate of Tippoo's Tiger: Defining Cultures through Public Display," *Cultural Encounters: Representing Otherness*, ed. Elizabeth Hallam and Brian V. Street (London: Routledge, 2000), p. 194.

23. Yash Shah, "History of Lathe from Beginning of Machine Tool Invention," *Yash Machine Tools* (blog), May 10, 2013, http://www.yashmachine.com/blog/history-of-lathe-from-beginning-of-machine-tool-invention (accessed March 20, 2018).

24. Beverly Steitz, "A Brief Computer History," Boston University, 2006, http://people.bu.edu/baws/brief%20computer%20history.html (accessed March 20, 2018).

25. Ibid.

CHAPTER 3. ROBOTICS: REDEFINING THE WORKFORCE

1. US Bureau of Labor Statistics, "Employer-Reported Workplace Injury and Illness Summary," news release, October 27, 2016, https://www.bls.gov/news.release/archives/osh_10272016.pdf (accessed March 20, 2018).

2. David Marshall, "Manufacturers Employ Robots to Improve Worker Health and Safety," ABB Conversations, November 16, 2012, https://www.abb-conversations.com/2012/11/manufacturers-employ-robots-to-improve-worker-health-and-safety (accessed March 20, 2018).

3. June Javelosa, "This Company's Productivity Soared after Replacing 90% of Employees with Robots," *Business Insider*, February 20, 2017, http://

www.businessinsider.com/companys-productivity-soared-after-replacing-90
-of-employees-with-robots-2017-2 (accessed March 20, 2018).

4. Andrew J. Oswald, Eugenio Proto, and Daniel Sgori, "Happiness and Productivity," *Journal of Labor Economics* 33, no. 4 (2015): 789–822, http:// wrap.warwick.ac.uk/63228/7/WRAP_Oswald_681096.pdf (accessed March 20, 2018).

5. Gary Robbins, "How Robots Will Change the American Workforce," *San Diego Union Tribune*, December 15, 2016, http://www.sandiego uniontribune.com/news/science/sd-me-robots-jobs-20161213-story.html (accessed March 20, 2018).

6. Ibid.

7. Harold L. Sirkin, Michael Zinser, and Justin Rose, "The Robotics Revolution: The Next Great Leap in Manufacturing," BCG Perspectives, September 23, 2015, https://www.bcg.com/publications/2015/lean -manufacturing-innovation-robotics-revolution-next-great-leap-manufacturing .aspx (accessed March 20, 2018).

8. Ibid.

9. Boston Consulting Group, "Global Spending on Robots Projected to Hit $87 Billion by 2025," press release, June 21, 2017, https://www.bcg .com/d/press/21june2017-gaining-robotics-advantage-162604 (accessed March 20, 2018).

10. International Federation of Robotics, "World Economics Report: European Union Occupies Top Position in the Global Automation Race," press release, September 29, 2016, https://ifr.org/ifr-press-releases/news/ world-robotics-report-2016 (accessed March 20, 2018).

11. Juha Heikkilä, "Robotics and Artificial Intelligence (Unit A.1)," European Commission, https://ec.europa.eu/digital-single-market/en/ content/robotics-and-artificial-intelligence-unit-a1 (accessed October 1, 2017).

12. Ibid.

13. Daron Acemoglu and Pascual Restrepo, "Robots and Jobs: Evidence from US Labor Markets," (working paper, National Bureau of Economic Research, Washington, DC, March 2017), http://www.nber.org/papers/ w23285.pdf (accessed March 20, 2018).

14. Carl Benedikt Frey and Michael Osborne, "The Future of Employment: How Susceptible Are Jobs to Computerization?" (working paper, Oxford Martin Programme on Technology and Employment, Oxford, UK, September 17, 2013), ttps://www.oxfordmartin.ox.ac.uk/publications/view/1314 (accessed March 20, 2018).

15. *The Future of Jobs: Employment, Skills, and Workforce Strategy for the Fourth Industrial Revolution* (Geneva, Switzerland: World Economic Forum, January 2016), http://reports.weforum.org/future-of-jobs-2016/ (accessed March 20, 2018).

16. Daron Acemoglu and Pascual Restrepo, "Robots and Jobs: Evidence from US Labor Markets" (working paper, National Bureau of Economic Research, Washington, DC, March 2017), http://www.nber.org/papers/w23285 (accessed March 20, 2018).

17. Mark Muro, "Where the Robots Are," *Brookings* (blog), August 14, 2017, https://www.brookings.edu/blog/the-avenue/2017/08/14/where-the-robots-are/?ex_cid=SigDig (accessed March 20, 2018).

18. Anna Hensel, "Brookings: Midwest and Southeast Employ the Most Robots," VentureBeat, August 14, 2017, https://venturebeat.com/2017/08/14/brookings-midwest-and-southeast-employ-the-most-robots (accessed March 20, 2018).

19. Muro, "Where the Robots Are."

20. Jonha Revesencio, "Why Happy Employees Are 12% More Productive," *Fast Company*, July 22, 2015, https://www.fastcompany.com/3048751/happy-employees-are-12-more-productive-at-work (accessed March 20, 2018).

21. Michael Addady, "Study: Being Happy at Work Really Makes You More Productive," *Fortune*, October 29, 2015, http://fortune.com/2015/10/29/happy-productivity-work (accessed March 20, 2018).

22. Revesencio, "Why Happy Employees."

23. *The World's Cities in 2016* (New York: United Nations, Department of Economic and Social Affairs, Population Division, 2016), http://www.un.org/en/development/desa/population/publications/pdf/urbanization/the_worlds_cities_in_2016_data_booklet.pdf (accessed March 20, 2018).

24. *Global Trends 2030: Alternative Worlds* (Washington, DC: US Office

of the Director of National Intelligence, December 2012), https://www.dni
.gov/files/documents/GlobalTrends_2030.pdf (accessed March 20, 2018).

CHAPTER 4. ARTIFICIAL INTELLIGENCE: FROM SCIENCE FICTION TO REALITY

1. Avery Thompson, "Chinese Scientists Successfully Teleported a Particle to Space," *Popular Mechanics*, July 12, 2017, http://www.popular mechanics.com/science/news/a27271/chinese-scientists-successfully -teleported-a-particle-to-space (accessed March 20, 2018).

2. Gene J. Koprowski, "Invisible Airplanes: Chinese, US Race for Cloaking Tech," Fox News, December 17, 2013, http://www.foxnews.com/ tech/2013/12/17/invisible-airplanes-chinese-us-scramble-for-cloaking-tech .html (accessed March 20, 2018).

3. Imperial College London, "Scientists Create First Working Invisibility Cloak," news release, October 19, 2006, https://www.imperial .ac.uk/news/2812/scientists-create-first-working-invisibility-cloak/ (accessed April 16, 2018).

4. Imperial College London, "'Space-Time Cloak' to Conceal Events Revealed in New Study," news release, November 16, 2010, https://www .imperial.ac.uk/news/94422/space-time-cloak-conceal-events-revealed-study/ (accessed April 16, 2018).

5. Luke Edwards, "Cloaking Devices Are Real, Here Are the Best Science Has Created," Pocket-lint, September 18, 2015, http://www.pocket -lint.com/news/131156-cloaking-devices-are-real-here-are-the-best-science -has-created (accessed March 20, 2018).

6. Mariella Moon, "Dubai Tests a Passenger Drone for Its Flying Taxi Service," Engadget, September 26, 2017, https://www.engadget.com/2017/ 09/26/dubai-volocopter-passenger-drone-test (accessed March 20, 2018).

7. Mark Vandevelde, "Amazon Is Creating More Jobs Than It Destroys," *Financial Times*, July 29, 2017, https://www.ft.com/content/cf98680c-738f -11e7-aca6-c6bd07df1a3c (accessed March 20, 2018).

8. Mark Molloy, "'Robot Lawyer' That Overturned 160,000 Parking

Tickets Now Helping Refugees," *Telegraph*, March 7, 2017, http://www
.telegraph.co.uk/technology/2017/03/07/robot-lawyer-overturned-160000
-parking-tickets-now-helping-refugees (accessed March 20, 2018); Rob Price,
"A Facebook Chatbot That Fought 250,000 Parking Fines Is Helping Refugees
Claim Asylum," *Business Insider*, March 6, 2017, http://uk.businessinsider
.com/facebook-chatbot-donotpay-help-refugees-claim-asylum-us-canada-uk
-joshua-browder-parking-fines-2017-3 (accessed March 20, 2017).

    9. Dan Mangan, "Lawyers Could Be the Next Profession to Be Replaced
by Computers," CNBC, February 17, 2017, https://www.cnbc
.com/2017/02/17/lawyers-could-be-replaced-by-artificial-intelligence.html
(accessed March 20, 2018).

    10. Ibid.

    11. Cecille De Jesus, "AI Lawyer 'Ross' Has Been Hired by Its First
Official Law Firm," Futurism, May 11, 2016, https://futurism.com/artificially
-intelligent-lawyer-ross-hired-first-official-law-firm (accessed March 20, 2018).

    12. Bona Benjamin, "Proceedings of a Summit on Preventing Patient
Harm and Death from I.V. Medication Errors," *American Journal of
Health-System Pharmacy* 65 (December 15, 2008), http://www.ajhp.org/
content/65/24/2367 (accessed March 20, 2018).

    13. Tim Dall et al., *The Complexities of Physician Supply and Demand:
Projections from 2013 to 2025* (Washington, DC: Association of American
Medical Colleges, March 2015), https://www.aamc.org/download/426242/
data/ihsreportdownload.pdf?cm_mmc=AAMC-_-ScientificAffairs-_-PDF
-_-ihsreport (accessed March 20, 2018).

    14. Tyler Durden, "Meet SAM, Brick Laying Robot That Does the Work
of 6 Humans," *Zero Hedge* (blog), March 28, 2017, http://www.zerohedge
.com/news/2017-03-27/meet-sam-brick-laying-robot-does-work-6-humans
(accessed March 20, 2018).

    15. Matt Simon, "This Robot Tractor Is Ready to Disrupt Construction,"
Wired, October 19, 2017, https://www.wired.com/story/this-robot-tractor-is
-ready-to-disrupt-construction (accessed March 20, 2018).

    16. April Glaser and Rani Molla, "The Construction Industry Is Short
on Human Workers and Ripe for a Robotic Takeover," Recode, June 6, 2017,
https://www.recode.net/2017/6/6/15701186/robots-construction-homes

-technology-drones-building-automation-productivity (accessed March 20, 2018).

17. Margi Murphy, "'Westworld'-Style Robots Could Live among Us by 2027," *New York Post*, May 22, 2017, https://nypost.com/2017/05/22/westworld-style-robots-could-live-among-us-by-2027/ (accessed March 20, 2018).

18. Pam Kragen, "World's First Talking Sex Robot Is Ready for Her Close-Up," *San Diego Union Tribune*, September 13, 2017, http://www.sandiegouniontribune.com/communities/north-county/sd-me-harmony-doll-20170913-story.html (accessed March 20, 2018).

19. Andy Haldane, "Labour's Share – Speech by Andy Haldane" (speech; London: Trades Union Congress, November 12, 2015), https://www.bankofengland.co.uk/speech/2015/labours-share (accessed March 20, 2018).

20. Ibid.

CHAPTER 5. CHALLENGES TO CONSIDER

1. Andy Haldane, "Labour's Share – Speech by Andy Haldane" (speech; London: Trades Union Congress, November 12, 2015), http://www.bankofengland.co.uk/publications/Pages/speeches/2015/864.aspx (accessed March 20, 2018).

2. Forrester, "Robots, AI Will Replace 7% of US Jobs by 2025," press release, June 22, 2016, https://www.forrester.com/Robots+AI+Will+Replace+7+Of+US+Jobs+By+2025/-/E-PRE9246 (accessed March 20, 2018).

3. Kristin Lee, "Artificial Intelligence, Automation, and the Economy," *Obama White House Archives* (blog), December 20, 2016, https://obamawhitehouse.archives.gov/blog/2016/12/20/artificial-intelligence-automation-and-economy (accessed March 20, 2018).

4. Grace Lordan and David Neumark, "People versus Machines: The Impact of Minimum Wages on Automatable Jobs" (working paper, National Bureau of Economic Research, Washington, DC, August 2017), http://www.nber.org/papers/w23667.pdf (accessed March 20, 2018).

5. Hayley Peterson, "McDonald's Shoots Down Fears It Is Planning to

Replace Cashiers with Kiosks," *Business Insider*, August 6, 2015, http://www
.businessinsider.com/what-self-serve-kiosks-at-mcdonalds-mean-for
-cashiers-2015-8 (accessed March 20, 2018).

6. Michael Hiltzik, "Does Andy Puzder Really Want to Replace His
Carl's Jr. Workers with Robots? No, But . . .," *Los Angeles Times*, March 30,
2016, http://www.latimes.com/business/hiltzik/la-fi-hiltzik-puzder
-20160322-snap-htmlstory.html (accessed March 20, 2018).

7. Shan Li, "Wendy's Adds Automation to the Fast-Food Menu," *Los
Angeles Times*, February 28, 2017, http://www.latimes.com/business/la-fi
-wendys-kiosk-20170227-story.html (accessed March 20, 2018).

8. Canadian Press, "Ontario Minimum Wage Hike Threatens 50,000
Jobs: Watchdog," *Toronto Sun*, September 12, 2017, http://www.torontosun
.com/2017/09/12/ontario-minimum-wage-hike-threatens-50000-jobs
-watchdog (accessed March 20, 2018).

9. "Ontario Minimum Wage Hike Sparks Debate Over Replacing
Human Workers with Machines," CBC News, June 14, 2017, http://www
.cbc.ca/news/canada/windsor/ontario-minimum-wage-hike-sparks-debate
-over-replacing-human-workers-with-machines-1.4160763 (accessed March
20, 2018).

10. Ibid.

11. "5 College Degrees That Will Be Extinct in 20 Years," WomensArticle,
December 8, 2016, http://www.womensarticle.com/5-college-degrees-that
-may-be-extinct-in-20-years/ (accessed October 1, 2017).

12. "What Is Corporate Personhood?" HRZone, https://www.hrzone
.com/hr-glossary/what-is-corporate-personhood (accessed March 20, 2018).

13. Kent Greenfield and Adam Winkler, "The US Supreme Court's
Cultivation of Corporate Personhood," *Atlantic*, June 24, 2015, https://www.
theatlantic.com/politics/archive/2015/06/raisins-hotels-corporate
-personhood-supreme-court/396773 (accessed March 20, 2018).

14. "Court in Argentina Grants Basic Rights to Orangutan," BBC,
December 21, 2014, http://www.bbc.com/news/world-latin-america
-30571577 (accessed March 20, 2018).

15. "Whanganui River the First in the World to Be Given Legal Status as
a Person," *Newshub*, March 15, 2017, http://www.newshub.co.nz/home/new

-zealand/2017/03/whanganui-river-the-first-in-the-world-to-be-given-legal
-status-as-a-person.html (accessed March 20, 2018).

16. European Parliament, "Robots: Legal Affairs Committee Calls for
EU-Wide Rules," press release, January 1, 2017, http://www.europarl.europa
.eu/news/en/press-room/20170110IPR57613/robots-legal-affairs
-committee-calls-for-eu-wide-rules (accessed March 20, 2018).

17. May Bulman, "EU to Vote on Declaring Robots to Be 'Electronic
Persons,'" *Independent*, January 13, 2017, http://www.independent.co.uk/life
-style/gadgets-and-tech/robots-eu-vote-electronic-persons-european-union-ai
-artificial-intelligence-a7527106.html (accessed March 20, 2018).

18. Markus Häuser, "Do Robots Have Rights? The European Parliament
Addresses Artificial Intelligence and Robotics," CMS Law, April 6, 2017,
http://www.cms-lawnow.com/ealerts/2017/04/do-robots-have-rights-the
-european-parliament-addresses-artificial-intelligence-and-robotics (accessed
March 20, 2018).

19. "Saudi Arabia Grants Citizenship to Humanoid Robot," *Russia Today*,
October 26, 2017, https://www.rt.com/news/407825-saudi-robot-citizen
-sophia (accessed March 20, 2018).

20. Kyle Wiggers, "Meet the 400-Pound Robots That Will Soon Patrol
Parking Lots, Offices, and Malls," Digital Trends, April 27, 2017, https://www
.digitaltrends.com/cool-tech/knightscope-robots-interview (accessed March
20, 2018).

21. Hassan Abbas, "Are Drone Strikes Killing Terrorists or Creating
Them?," *Atlantic*, March 31, 2013, https://www.theatlantic.com/
international/archive/2013/03/are-drone-strikes-killing-terrorists-or-creating
-them/274499 (accessed March 20, 2018).

22. "Military Robots," Robots and Androids, http://www.robots-and
-androids.com/military-robots.html (accessed October 1, 2017).

23. J. R. Potts, "MULE (Multifunction Utility / Logistics and Equipment) Unmanned Infantry Support Vehicle (UISV)," Military Factory, last
modified June 2, 2016, https://www.militaryfactory.com/armor/detail.asp
?armor_id=314 (accessed March 20, 2018).

24. "Military Robots."

25. Dave Gershgorn, "Police Used Bomb Disposal Robot to Kill a Dallas

Shooting Suspect," *Popular Science*, July 8, 2016, https://www.popsci.com/police-used-bomb-disposal-robot-to-kill-dallas-shooting-suspect (accessed March 20, 2018).

26. Patrick Tucker, "Armed Ground Robots Could Join the Ukrainian Conflict Next Year," Defense One, October 10, 2017, http://www.defenseone.com/technology/2017/10/armed-ground-robots-could-make-their-combat-debut-ukrainian-conflict-next-year/141677 (accessed March 20, 2018).

27. Emily Feng, "China Agency Targets High-Tech Weapons Development," *Financial Times*, July 26, 2017, https://www.ft.com/content/2c9b4370-71c5-11e7-aca6-c6bd07df1a3c (accessed March 20, 2018).

28. "'Whoever Leads in AI Will Rule the World': Putin to Russian Children on Knowledge Day," *Russian Times*, September 1, 2017, https://www.rt.com/news/401731-ai-rule-world-putin (accessed March 20, 2018).

29. Jamie Micklethwaite, "Sex Robot ARMIES: Fears Hackers Could Create Killer Cyborgs and Turn Technology on Punters," *Daily Star*, September 9, 2017, http://www.dailystar.co.uk/news/latest-news/643302/sex-robots-hackers-killer-cyborgs-technology-elon-musk-artificial-intelligence-world-war-3 (accessed March 20, 2018).

30. "Rogue Robots: Testing the Limits of an Industrial Robot's Security," Trend Micro, May 3, 2017, https://www.trendmicro.com/vinfo/us/security/news/internet-of-things/rogue-robots-testing-industrial-robot-security (accessed March 20, 2018).

31. Andy Greenberg, "How Hackable Is Your Car? Consult This Handy Chart," *Wired*, August 6, 2014, https://www.wired.com/2014/08/car-hacking-chart (accessed March 20, 2018).

32. Mohit Kumar, "Unpatchable Flaw in Modern Cars Allows Hackers to Disable Safety Features," *Hacker News*, August 17, 2017, https://thehackernews.com/2017/08/car-safety-hacking.html (accessed March 20, 2018).

33. "Julian Assange Reveals the CIA's Biggest Secret: The Weeping Angel," Digital News Network, March 8, 2017, http://www.digitalnewsnetwork.net/2017/03/08/julian-assange-reveals-the-cias-biggest-secret-the-weeping-angel (accessed March 20, 2018).

34. Patrick Tucker, "The Future the US Military Is Constructing: A Giant, Armed Nervous System," Defense One, September 26, 2017, http://

www.defenseone.com/technology/2017/09/future-us-military-constructing
-giant-armed-nervous-system/141303 (accessed March 20, 2018).

35. Mark Austin, "Elon Musk Is Convinced Killer Robots Are Coming, and He Has a Plan," Digital Trends, August 20, 2017, https://www
.digitaltrends.com/cool-tech/ban-killer-robots (accessed March 20, 2018).

36. "Global Race for AI Will 'Most Likely Cause' WWIII as Computers Launch 1ˢᵗ Strike – Musk," *Russia Today*, September 4, 2017, https://www
.rt.com/usa/401957-ww3-ai-musk-strike (accessed March 20, 2018).

37. US Department of Defense, "Contracts," press release, May 31, 2017, https://www.defense.gov/News/Contracts/Contract-View/Article/1198370
(accessed March 20, 2018).

38. Jon Lockett, "US Military Will Have More Combat Robots Than Human Soldiers by 2025," *New York Post*, June 15, 2017, http://nypost
.com/2017/06/15/us-military-will-have-more-combat-robots-than-human
-soldiers-by-2025 (accessed March 20, 2018).

CHAPTER 6. DISRUPTION, EXPONENTIAL THINKING, AND TIPPING POINTS

1. Andy Rachleff, "What 'Disrupt' Really Means," *TechCrunch*, February 16, 2013, https://techcrunch.com/2013/02/16/the-truth-about-disruption
(accessed March 20, 2018).

2. According to Investopedia, "A special purpose vehicle/entity (SPV/ SPE) is a subsidiary company with an asset/liability structure and legal status that makes its obligations secure even if the parent company goes bankrupt." Investopedia, s v "Special Purpose Vehicle/Entity - SPV/SPE," https://www
.investopedia.com/terms/s/spv asp (accessed April 16, 2018).

3. Kurt Eichenwald, "Enron's Collapse; Audacious Climb to Success Ended in a Dizzying Plunge," *New York Times*, January 13, 2002, http://www
.nytimes.com/2002/01/13/us/enron-s-collapse-audacious-climb-to-success
-ended-in-a-dizzying-plunge.html (accessed March 20, 2018).

4. Carl Gutierrez, "Bear Stearns Announces More Job Cuts," *Forbes*, October 3, 2007, https://www.forbes.com/2007/10/03/bear-stearns-layoffs

-markets-equity-cx_cg_1003markets23.html#31470483395d (accessed March 20, 2018).

5. "An Exponential Primer: Your Guide to Our Essential Concepts," Singularity University, https://su.org/concepts (accessed October 1, 2017).

6. Gordon E. Moore, "Cramming More Components onto Integrated Circuits," *Electronics* (April 19, 1965): 114–17, http://www.cs.utexas.edu/~fussell/courses/cs352h/papers/moore.pdf (accessed March 20, 2018).

7. "Peter H. Diamandis, MD," XPRIZE, https://www.xprize.org/about/board-of-directors/peter-h-diamandis-md (accessed October 1, 2017).

8. Peter H. Diamandis, "The Difference between Linear and Exponential Thinking," Big Think, 2017, http://bigthink.com/in-their-own-words/the-difference-between-linear-and-exponential-thinking#videos-nav-dropdown-68 (accessed October 1, 2017).

9. "Cascade Effect Thinking," Institute for Cascade Effect Research, https://www.cascadeeffects.com/cascade-effect-thinking.html (accessed October 1, 2017).

10. *Merriam-Webster*, s.v. "Tipping Point," last updated February 23, 2018, https://www.merriam-webster.com/dictionary/tipping%20point (accessed March 20, 2018).

11. Stuart Taylor, "Looking Back at 2015: A Tipping Point for the Internet of Things," *Cisco* (blog), January 5, 2016, https://blogs.cisco.com/sp/looking-back-at-2015-a-tipping-point-for-the-internet-of-things (accessed March 20, 2018).

12. Daniel Lin, "How a Pacific Island Changed from Diesel to 100% Solar Power," *National Geographic*, February 23, 2017, https://news.nationalgeographic.com/2017/02/tau-american-samoa-solar-power-microgrid-tesla-solarcity (accessed March 20, 2018).

13. Jack Ewing, "Volvo, Betting on Electric, Moves to Phase Out Conventional Engines," *New York Times*, July 5, 2017, https://www.nytimes.com/2017/07/05/business/energy-environment/volvo-hybrid-electric-car.html (accessed March 20, 2018).

14. *Deep Shift: Technology Tipping Points and Societal Impact* (Geneva, Switzerland: World Economic Forum, September 2015), http://www3.weforum.org/docs/WEF_GAC15_Technological_Tipping_Points_report_2015.pdf (accessed March 20, 2018).

15. Michael Chui, James Manyika, and Mehdi Miremadi, "Four Fundamentals of Workplace Automation," *McKinsey Quarterly*, November 2015, https://www.mckinsey.com/business-functions/digital-mckinsey/our -insights/four-fundamentals-of-workplace-automation (accessed March 20, 2018).

16. "First Smart Pharmacy Run by Robot Begins at Rashid Hospital," *Gulf News*, January 13, 2017, http://gulfnews.com/news/uae/health/first -smart-pharmacy-run-by-robot-begins-at-rashid-hospital-1.1961228 (accessed March 20, 2018).

17. Mark Bain, "Forget Wearables. In the Future, Your Clothes Will Connect to the Internet," Quartz Media, November 20, 2016, https:// qz.com/829521/forget-wearables-in-the-future-your-clothes-will-connect-to -the-internet (accessed March 20, 2018).

18. MarEx, "US Navy Researchers 3D-Print a Small Submarine," *Maritime Executive*, July 31, 2017, https://www.maritime-executive.com/article/ us-navy-researchers-3d-print-a-small-submarine (accessed March 20, 2018).

19. James Manyika et al., "Disruptive Technologies: Advances That Will Transform Life, Business, and the Global Economy," *McKinsey Quarterly*, May 2013, https://www.mckinsey.com/business-functions/digital-mckinsey/our -insights/disruptive-technologies (accessed March 20, 2018).

CHAPTER 7. TECHNOLOGICAL AND ECONOMIC SINGULARITY

1. Vernor Vinge, "The Coming Technological Singularity: How to Survive in the Post-Human Era" (prepared for the VISION-21 Symposium, NASA Lewis Research Center and the Ohio Aerospace Institute, Westlake, OH, March 31, 1993), https://edoras.sdsu.edu/~vinge/misc/singularity.html (accessed March 20, 2018).

2. Don Galeon and Christianna Reedy, "Kurzweil Claims That the Singularity Will Happen by 2045," Futurism, October 5, 2017, https:// futurism.com/kurzweil-claims-that-the-singularity-will-happen-by-2045 (accessed March 20, 2018).

3. Don Galeon and Kristin Houser, "Softbank CEO: The Singularity Will Happen by 2047," Futurism, March 1, 2017, https://futurism.com/softbank-ceo-the-singularity-will-happen-by-2047 (accessed March 20, 2018).

4. Investopedia, s.v. "Moore's Law," http://www.investopedia.com/terms/m/mooreslaw.asp (accessed October 1, 2017).

5. Siobhan McFadyen, "Eurozone Turmoil: Spain, Italy, and Greece Owe Massive Debt of €1 Trillion to ECB," Express, March 20, 2017, http://www.express.co.uk/news/world/781518/Eurozone-finance-economy-trillion-debt-ECB-Spain-Italy-Portugal-Greece (accessed March 20, 2018).

6. Bob Bryan, "The US Government Just Passed $20 Trillion in Debt for the First Time Ever," Business Insider, September 11, 2017, http://www.businessinsider.com/us-debt-20-trillion-how-much-2017-9 (accessed March 20, 2018).

7. Kimberly Amadeo, "How Interest on the National Debt Affects You," Balance, July 6, 2017, https://www.thebalance.com/interest-on-the-national-debt-4119024 (accessed March 20, 2018).

8. Mike Collins, "The Big Bank Bailout," Forbes, July 14, 2015, https://www.forbes.com/sites/mikecollins/2015/07/14/the-big-bank-bailout/#71eae5f22d83 (accessed March 20, 2018).

9. Marjie Bloy, "The Luddites 1811–1816," Victorian Web, last modified December 30, 2005, http://www.victorianweb.org/history/riots/luddites.html (accessed March 20, 2018).

CHAPTER 8. THE AGE OF THE 'PRENEUR

1. Rachel Hallett and Rosamond Hutt, "10 Jobs That Didn't Exist 10 Years Ago," World Economic Forum, June 7, 2016, https://www.weforum.org/agenda/2016/06/10-jobs-that-didn-t-exist-10-years-ago/ (accessed March 20, 2018).

2. "Fastest Growing Occupations," Occupational Outlook Handbook, US Bureau of Labor Statistics, last modified October 24, 2017, https://www.bls.gov/ooh/fastest-growing.htm (accessed March 20, 2017).

3. James Manyika et al., "Jobs Lost, Jobs Gained: Workforce Transitions

in a Time of Automation," McKinsey Global Institute, November 2017, https://www.mckinsey.com/global-themes/future-of-organizations-and -work/what-the-future-of-work-will-mean-for-jobs-skills-and-wages (accessed March 20, 2018).

4. Thomas R. Eisenmann, "Entrepreneurship: A Working Definition," *Harvard Business Review*, January 10, 2013, https://hbr.org/2013/01/what-is -entrepreneurship (accessed March 20, 2018).

5. Ibid.

6. Charles A. Jeszeck, *Contingent Workforce: Size, Characteristics, Earnings, and Benefits* (Washington, DC: US Government Accountability Office, April 20, 2015), https://www.gao.gov/products/GAO-15-168R (accessed March 20, 2018).

7. "About Us," Upwork, https://www.upwork.com/about (accessed October 1, 2017).

8. Upwork, "Freelancers Union and Upwork Release New Study Revealing Insights into the Almost 54 Million People Freelancing in America," press release, https://www.upwork.com/press/2015/10/01/freelancers-union -and-upwork-release-new-study-revealing-insights-into-the-almost-54-million -people-freelancing-in-america (accessed October 1, 2017).

9. Claire Zillman, "PwC Wants to Use 'Gig Economy' Workers to Staff Projects for Its Clients," *Fortune*, March 7, 2016, http://fortune.com/ 2016/03/07/pwc-freelance-marketplace (accessed March 20, 2018).

10. "Chan Zuckerberg Initiative," *Chan Zuckerberg*, https://chan zuckerberg.com (accessed October 1, 2017).

11. Ibid.

12. "Personal Branding Guru, William Arruda," YouTube video, 4:11, posted by "William Arruda," May 5, 2009, https://youtu.be/6paItEm2AF4?t =1m10s (accessed March 20, 2018).

13. Jesko Perrey, Tjark Freundt, and Dennis Spillecke, "The Brand Is Back: Staying Relevant in an Accelerating Age," *McKinsey Quarterly*, May 2015, https://www.mckinsey.com/business-functions/marketing-and-sales/ our-insights/the-brand-is-back-staying-relevant-in-an-accelerating-age (accessed March 20, 2018).

14. Lee W. Frederiksen, "The Research behind Why You Should Grow

Your Personal Brand," *International Coach Federation* (blog), November 4, 2014, http://coachfederation.org/blog/index.php/3581 (accessed March 20, 2018).

15. "Values First! Personal Values Assessment," Success Rockets, http://valuesfirstassessment.com.

16. *Workplace Conflict and How Businesses Can Harness It to Thrive* (Sunnyvale, CA: CPP, July 2008), http://img.en25.com/Web/CPP/Conflict_report.pdf (accessed March 21, 2018).

17. Andrew J. Oswald, "Happiness and Economic Performance," (paper, Department of Economics, Warwick University, Coventry, England, April 1997), https://warwick.ac.uk/fac/soc/economics/staff/ajoswald/happecperf.pdf (accessed March 21, 2018).

18. Kelley Holland, "In Mission Statement, Bizspeak and Bromides," *New York Times*, September 23, 2007, http://www.nytimes.com/2007/09/23/jobs/23mgmt.html (accessed March 21, 2018).

19. "Steve Jobs Unveils the Think Different Campaign," YouTube video, 16:01, posted by "EverySteveJobsVideo," March 17, 2015, https://www.youtube.com/watch?v=4HsGAc0_Y5c (accessed March 21, 2018).

20. Ibid.

21. Dan Schawbel, "Hire for Attitude," *Forbes*, January 23, 2012, http://www.forbes.com/sites/danschawbel/2012/01/23/89-of-new-hires-fail-because-of-their-attitude (accessed March 21, 2018).

22. Daniel Goleman, "75 Years Later, Study Still Tracking Geniuses," *New York Times*, March 7, 1995, http://www.nytimes.com/1995/03/07/science/75-years-later-study-still-tracking-geniuses.html?pagewanted=all (accessed March 21, 2018).

23. Geoff Williams, "Here's What Banking and Money Will Be Like 30 Years from Now," *Business Insider*, September 20, 2013, http://www.businessinsider.com/heres-what-banking-and-money-will-be-like-30-years-from-now-2013-9 (accessed March 21, 2018).

24. Larry Boyer, "Should You Be Worried about AI Taking Your Job Away?," *Forbes*, July 10, 2017, https://www.forbes.com/sites/forbescoaches council/2017/07/10/should-you-be-worried-about-ai-taking-your-job-away/#1b13de3d4843 (accessed March 21, 2018).

CHAPTER 9. REMAINING FINANCIALLY SOUND IN DISRUPTIVE TIMES

1. Garrett B. Gunderson with Stephen Palmer, *Killing Sacred Cows: Overcome the Financial Myths That Are Destroying Your Prosperity* (Austin, TX: Greenleaf, 2008).

2. "Crowdfunding," US Securities and Exchange Commission, last modified November 2, 2016, https://www.sec.gov/spotlight/crowdfunding .shtml (accessed November 28, 2017).

3. "What Is Crowdfunding?," Fundable, https://www.fundable.com/ learn/resources/guides/crowdfunding-guide/what-is-crowdfunding (accessed November 28, 2017).

4. "Investor Bulletin: Regulation A," US Securities and Exchange Commission, July 8, 2015, last modified February 6, 2017, https://www.sec .gov/oiea/investor-alerts-bulletins/ib_regulationa.html (accessed November 28, 2017).

5. "Investor Bulletin: Initial Coin Offerings," US Securities and Exchange Commission, July 25, 2017, https://www.sec.gov/oiea/investor -alerts-and-bulletins/ib_coinofferings (accessed November 28, 2017).

6. Karen Rands, *Inside Secrets to Angel Investing: Step-by-Step Strategies to Leverage Private Equity Investment for Passive Wealth Creation* (Business Publications, 2017).

7. Investopedia, s.v. "Angel Investor," https://www.investopedia.com/ terms/a/angelinvestor.asp (accessed November 28, 2017).

8. Larry Boyer, "Can You Be an Angel Investor? The Answer May Surprise You," LinkedIn, August 7, 2017, https://www.linkedin.com/pulse/can-you -angel-investor-answer-may-surprise-larry-boyer/ (accessed March 21, 2018).

9. "1-2-3 Success! The Reach Personal Branding Process," Reach Personal Branding, 2009, https://www.reachcc.com/reachdotcom.nsf/bdf8f 1dec3dadac0c1256aa700820c2c/7fc445a72ce0dbe5c1256af500027f28!Open Document (accessed November 27, 2017).

10. "Why Connecting Is Not Enough," YouTube video, 5:41, posted by "Andy Lopata," March 21, 2014, https://www.youtube.com/watch?v=IKud _lWw2rY (accessed March 21, 2018).

CHAPTER 10. GETTING OVER IT AND GETTING STARTED

1. Alex Santoso, "Four Geeky Laws That Rule Our World," *Neatorama* (blog), September 5, 2012, http://www.neatorama.com/2012/09/05/Four -Geeky-Laws-That-Rule-Our-World/ (accessed March 21, 2018).

2. Dan Schawbel, *Promote Yourself: The New Rules for Career Success* (New York: St. Martin's, 2013), p. 27.

3. "Table 7. Survival of Private Sector Establishments by Opening Year" (Washington, DC: US Bureau of Labor Statistics, 2016), https://www.bls .gov/bdm/us_age_naics_00_table7.txt (accessed March 21, 2018).

4. Greg S. Reid, *Stickability: The Power of Perseverance* (New York: Jeremy P. Tarcher/Penguin, 2013).

5. Napoleon Hill, *Think and Grow Rich* (Virginia: Napoleon Hill Foundation, 1960), pp. 43–44; Sharon L. Lechter and Greg S. Reid, *Three Feet from Gold: Turn Your Obstacles into Opportunities* (New York: Sterling, 2009), pp. 21–25.

FINAL THOUGHTS

1. Quentin Hardy, "Gearing Up for the Cloud, AT&T Tells Its Workers: Adapt, or Else," *New York Times*, February 13, 2016, https://www.nytimes .com/2016/02/14/technology/gearing-up-for-the-cloud-att-tells-its-workers -adapt-or-else.html (accessed March 21, 2018).

2. Laura Noonan, Patrick Jenkins, and Olaf Storbeck, "Deutsche Bank Chief Hints at Thousands of Job Losses," *Financial Times*, November 8, 2017, https://www.ft.com/content/e7844048-c3e5-11e7-a1d2-6786f39ef675 (accessed March 21, 2018).

3. Larry Boyer, "Should You Listen to What CEOs Are Saying?" LinkedIn, February 18, 2016, https://www.linkedin.com/pulse/should-you -listen-what-ceos-saying-larry-boyer/ (accessed March 21, 2018).

4. Tibi Puiu, "The AI Buster-Buster: New Machine Dominates AI That Dominated Humans at GO," ZME Science, last modified October 19, 2017,

https://www.zmescience.com/science/news-science/alphago-zero
-unstoppable-432432 (accessed March 21, 2018).

    5. Ibid.

SELECTED BIBLIOGRAPHY

Arruda, William, and Kirsten Dixson. *Career Distinction: Stand Out by Building Your Brand.* Hoboken, NJ: John Wiley & Sons, 2007.

Bates, Laura. "The Trouble with Sex Robots." *New York Times*, July 17, 2017. https://www.nytimes.com/2017/07/17/opinion/sex-robots-consent.html (accessed December 1, 2017).

"The Effects of Optimism, Pessimism, and Anger on Health." *EHE Newsletter* 13, no. 22 (May 29, 2013). https://www.eheandme.com/news_articles/549652445 (accessed December 1, 2017).

Kurzweil, Ray. "We Are Entering the Singularity." Kurzweil Accelerating Intelligence, October 6, 2015. http://www.kurzweilai.net/celebrating-the-10-year-anniversary-of-book-the-singularity-is-near (accessed December 1, 2017).

Martin, James. "The 17 Great Challenges of the Twenty-First Century." Oxford University, January 2007. http://www.elon.edu/docs/e-web/predictions/17_great_challenges.pdf (accessed December 1, 2017).

Monitoring Environmental Progress: A Report on Work in Progress. Washington, DC: World Bank, September 30, 1995. http://documents.worldbank.org/curated/en/378701468765915443/Monitoring-environmental-progress-a-report-on-work-in-progress (accessed December 1, 2017).

Poole, Heather. "Job Creation by Startups and Young Companies." Connecticut General Assembly, 2016. https://www.cga.ct.gov/2016/rpt/2016-R-0003.htm (accessed December 1, 2017).

Rossi, Luke. "How Might the Rise of Robots Affect Financial Risk?" Barclay Simpson, January 22, 2016. http://www.barclaysimpson.com/news/how-might-the-rise-of-robots-affect-financial-risk-news-801810015 (accessed December 1, 2017).

Sinclair, Upton. *The Jungle.* New York: Doubleday, 1906.

INDEX